WELCOME TO YOUR BRAIN

WELCOME TO YOUR BRAIN

Why You Lose Your Car Keys but Never Forget How to Drive and Other Puzzles of Everyday Life

SANDRA AAMODT, Ph.D.
and SAM WANG, Ph.D.

BLOOMSBURY

Art credits:
Illustrations throughout by Lisa Haney. Page 5 courtesy of Michael MacAskill; page 17 (left) courtesy of Kenneth Catania; pages 17 (right), 19, 26, and 166 courtesy of Sam Wang; page 23 courtesy of Patrick Lane; page 43 courtesy of Edward Adelson, reprinted with permission; page 44 reprinted with permission from P. Thompson "Margaret Thatcher: A New Illusion," *Perception* 9 (1980): 483–84.

Published by Bloomsbury USA, New York
Distributed to the trade by Macmillan

All papers used by Bloomsbury USA are natural, recyclable products made from wood grown in well-managed forests. The manufacturing processes conform to the environmental regulations of the country of origin.

LIBRARY OF CONGRESS CATALOGING-IN-PUBLICATION DATA

Aamodt, Sandra.
Welome to your brain: why you lose your car keys but never forget how to drive and other puzzles of everyday life / Sandra Aamodt and Sam Wang.—1st U.S. ed.
p. cm.
ISBN-13: 978-1-59691-283-0
ISBN-10: 1-59691-283-9
1. Brain—Popular works. 2. Neurophysiology—Popular works. I. Wang, Sam, 1967– II. Title.

QP376.A222 2008
612.8'2—dc22 2007026739

First U.S. Edition 2008

1 2 3 4 5 6 7 8 9 10

Designed by Elizabeth Van Itallie
Printed in the United States of America by Quebecor World Fairfield

From Sandra, to Ken and Aquila
From Sam, to Dad, Becca, and Vita

Contents

Part 1 – Your Brain and the World

Part 2 – Coming to Your Senses

Part 3 – How Your Brain Changes Throughout Life

PART 5 – YOUR RATIONAL BRAIN

Part 6 – Your Brain in Altered States

ACKNOWLEDGMENTS

In our careers so far, we have written over half a million words about the brain, but that experience only partially prepared us for writing this book. We have wondered why acknowledgments run so long. Now we know.

When Jack Horne learned that both of us were planning to write the same book, he suggested we combine our efforts. Sandy Blakeslee and Jeff Hawkins recommended their agency, Levine Greenberg, to us, and vice versa. Our agent, Jim Levine, and his assistant, Lindsay Edgecombe, helped us shape the book's tone and content. All authors should have such expert guides for their first book. Beth Fisher connected us with publishers around the world. At Bloomsbury USA we have been lucky to work with our editor, Gillian Blake, who has been enthusiastic from the beginning and has provided an experienced hand. She, Ben Adams, and the Bloomsbury crew have improved our words and thoughts and kept us moving forward. Thanks are also due to Lisa Haney and Patrick Lane for beautiful illustrations and to Ken Catania, Pete Thompson, Ted Adelson, and Michael MacAskill for permission to use technical images.

We wrote a substantial part of the book at the Villa Serbelloni on the shores of Lake Como in Bellagio, Italy, an experience made possible by the Rockefeller Foundation and words of support from Jane Flint, Bob Horvitz, Charles Jennings, Olga Pellicer, Robert Sapolsky, and Shirley Tilghman. Pilar Palacia, Elena Ongania, and the rest of the Villa Serbelloni staff created an elegant but relaxed atmosphere for thinking, talking, and writing. Our fellow residents provided a great forum and we thank them all: Anne Waldman, Ed Bowes, Seemin Qayum, Sinclair Thomson, Raka Ray, Ashok Bardhan, Richard Cooper, Joan Kennelly, Jane Burbank, Fred Cooper, Russell Gordon, Jennifer Pierce, Dedre Gentner, Ken Forbus, David and Kathy Ringrose, Len and Gerry Pearlin, Bishakha Datta, Gautam Ojha, Sushil Sharma, Helen Roberts, Rodney Barker, Cyrus Cassells, Andrée Durieux-Smith, and Roger Smith.

Friends, colleagues, and students helped and encouraged us tremendously and were the source of invaluable suggestions, discussions, and corrections. We are especially grateful to Ralph Adolphs, Daphne Bavelier, Alim-Louis Benabid, Karen Bennett, Michael Berry, Ken Britten, Carlos Brody, Tom Carmichael, Gene Civillico, Mike DeWeese, David Eagleman, Neir Eshel, Michale Fee, Asif Ghazanfar, Mark Goldberg, Astrid Golomb, Liz Gould, David Grodberg, Patrick Hof, Hans Hofmann, Petr Janata, Danny Kahneman, Yevgenia Kozorovitskiy,

Ivan Kreilkamp, Eric London, Zach Mainen, Eve Marder, David Matthews, Becca Moss, Eric Nestler, Elissa Newport, Bill Newsome, Bob Newsome, Yael Niv, Liz Phelps, Robert Provine, Kerry Ressler, Rebecca Saxe, Clarence Schutt, Steven Schultz, Mike Schwartz, Mike Shadlen, Debra Speert, David Stern, Chess Stetson, Russ Swerdlow, Ed Tenner, Leslie Vosshall, Larry Young, and Gayle Wittenberg. Sam thanks his entire laboratory for accommodating his preoccupation, especially Rebecca Khaitman for excellent assistance. The Princeton University library was an essential resource. Finally, we thank Ivan Kaminow for telling us about the cell phone trick. Any remaining problems with the science, of course, are our responsibility and not theirs.

Our spouses went far beyond the call of duty in supporting us and this project, keeping us as sane as possible. Sandra thanks Ken Britten for his tolerant amusement at the prospect of entertaining himself for yet another weekend while she worked on the book and for his enthusiastic contributions to many shared adventures. She also thanks her parents, Roger and Jan Aamodt, for teaching her that girls too can take risks in pursuit of their dreams. Sam thanks Becca Moss for her partnership, her aplomb in the face of yet another crazy idea that got out of hand, and for providing a light when things got dark. Finally, Sam thanks his parents, Chia-lin and Mary Wang, for planting the seeds of a lifelong love of science and learning.

QUIZ

How Well Do You Know Your Brain?

Before you start reading this book, find out what you already know about your brain.

1) When are your last brain cells born?

(a) Before birth

(b) At age six

(c) Between the ages of eighteen and twenty-three

(d) In old age

2) Men and women have inborn differences in

(a) spatial reasoning

(b) strategies for navigation

(c) ability to leave the toilet seat down

(d) Both a and b

(e) Both b and c

3) Which of the following is *not* likely to improve brain function in old age?

(a) Eating fish with omega-3 fatty acids

(b) Getting regular exercise

(c) Drinking one or two glasses of red wine per day

(d) Drinking a whole bottle of red wine per day

4) Which of the following strategies is the best one for overcoming jet lag?

(a) Taking melatonin the night after you arrive at your destination

(b) Avoiding daylight for several days

(c) Getting sunlight in the afternoon at your destination

(d) Sleeping with the lights on

5) Your brain uses about as much energy as

(a) a refrigerator light

(b) a laptop computer

(c) an idling car

(d) a car moving down a freeway

6) Your friend is trying to tickle your belly. You can reduce the tickling sensation by

(a) putting your hand on his to follow the movement

(b) biting your knuckles

(c) tickling him back

(d) drinking a glass of water

7) Which of the following activities is likely to improve performance in school?

(a) Listening to classical music while you sleep

(b) Listening to classical music while you study

(c) Learning to play a musical instrument as a child

(d) Taking breaks from studying to play video games

(e) Both c and d

8) Which of the following things is a blow to the head least likely to cause?

(a) Loss of consciousness

(b) Memory loss

(c) Restoration of memory after suffering amnesia

(d) Personality change

9) Which of the following activities before a test might help you to perform better? (You may choose more than one.)

(a) Having a drink

(b) Having a cigarette

(c) Eating a candy bar

(d) Telling yourself with great conviction that you are good at this kind of test

10) You are in a noisy room, attempting to talk to your friend on your cell phone. To have a clearer conversation, you should

(a) talk more loudly

(b) cover one ear and listen through the other

(c) cover your ear when you talk

(d) cover the mouthpiece when you listen

11) Which of the following is an effective way to reduce anxiety?

(a) Antidepressant drugs

(b) Exercise

(c) Behavioral therapy

(d) All of the above

12) Which of the following is the hardest thing your brain does?

(a) Doing long division

(b) Looking at a photograph

(c) Playing chess

(d) Sleeping

13) Blind people are better than sighted people at which of the following?

(a) Understanding words

(b) Hearing sounds

(c) Remembering stories

(d) Training dogs

14) Your mother was improving your brain capacity when she told you which of the following things?

(a) "Turn that music down"

(b) "Go out and play"

(c) "Practice your instrument"

(d) All of the above

15) Memory starts to get worse in which decade of life?

(a) Thirties

(b) Forties

(c) Fifties

(d) Sixties

16) Which activities kill brain cells?

(a) Drinking three bottles of beer in an evening

(b) Smoking a joint

(c) Dropping acid

(d) All of the above

(e) None of the above

17) Which depiction of neurological damage is least realistic?

(a) Guy Pearce's character Leonard in *Memento*

(b) Drew Barrymore in *50 First Dates*

(c) Dory in *Finding Nemo*

(d) John Nash in *A Beautiful Mind*

18) What percentage of mammalian species are monogamous?

(a) 5%

(b) 25%

(c) 50%

(d) 90%

19) What percentage of your brain do you use?

(a) 10%

(b) 5% when you are sleeping, 20% when you are awake

(c) 100%

(d) Varies according to intelligence

20) When Einstein's brain was compared with the average person's, it

(a) was larger

(b) was indistinguishable in size

(c) had more folds on the surface

(d) had an extra part

Answers: 1) d, 2) d, 3) d, 4) c, 5) a, 6) a, 7) e, 8) c, 9) b and d, 10) d, 11) d, 12) b, 13) c, 14) d, 15) a, 16) e, 17) b, 18) a, 19) c, 20) b

INTRODUCTION

YOUR BRAIN: A USER'S GUIDE

I used to think my brain was my most important organ. But then I thought: wait a minute, who's telling me that?

—Emo Phillips

In our decades of studying and writing about neuroscience, we have often found ourselves discussing the brain in strange places: at the salon, in taxicabs, and even in the occasional elevator. Believe it or not, people don't run away (usually). Instead, they ask us all sorts of interesting questions: "When I drink, am I killing my brain cells?" "Does cramming for an exam work?" "Will playing music during pregnancy make my baby smarter?" "What is wrong with my teenager [or parent]?" "Why can't you tickle yourself?" "Do men and women think differently?" "Can you really get amnesia from being hit on the head?"

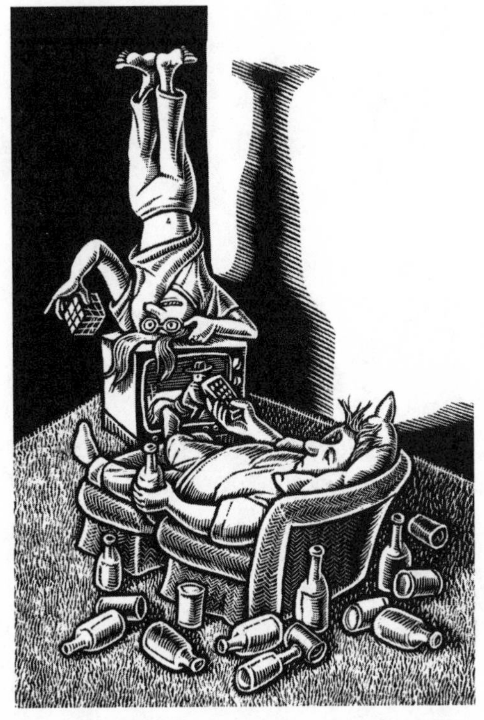

All these questions lead to your brain, the amazing three pounds in your skull that make you yourself. Your brain lets you watch a sunset, learn a language, tell a joke, recognize a friend, run from danger, and read this sentence.

In the last twenty years, neuroscientists have learned a lot about how your brain does all these things. It's a complicated subject, but we think it doesn't have to be intimidating. This book will give you the inside scoop on how your brain really works—and how you can help it work better.

Your brain has many ways of doing its job, including tricks and shortcuts that help it work efficiently—but may lead you to make predictable mistakes. By reading this book, you'll find out how you accomplish the things you do every day. Along the way we'll explode some of the myths that you might believe because "everybody knows" they're true. For instance, you don't really use only 10 percent of your brain. (Come on.)

Knowing your brain better can be both fun and useful. We will show you simple changes that will allow you to do more with your brain and help you lead a happier and more productive life. We'll also show you how disease can damage your brain—and suggest ways to prevent or repair this damage.

This book is like a guided tour: we'll see all the best sights and most important spots. But you don't have to start at the beginning. You can dip in anywhere and read this book in small pieces because each chapter stands on its own. In each one, you'll find fun facts, cocktail party–ready stories to amuse your friends, and practical tips to help you use your brain better.

- In part 1, we introduce the star of the show, your brain. We pull aside the curtain to show what is happening behind the scenes and explain how your brain helps you survive in the world.
- In part 2, we take a tour of your senses, explaining how you see, hear, touch, smell, and taste.
- In part 3, we show how your brain changes through life, from birth to old age.
- In part 4, we examine your brain's emotional systems, focusing on how they help you navigate life effectively.
- In part 5, we discuss your reasoning abilities, including decision making, intelligence, and gender differences in cognition.
- In part 6, we examine altered states of your brain—consciousness, sleep, drugs and alcohol, and disease.

Leave this book by your bedside or on your coffee table, and dip in anywhere, anytime. We hope you'll be enlightened and entertained, and that after reading a few pages you will want to read the whole book. Now pull up a chair and get ready to find out about your brain—and about yourself!

PART ONE

YOUR BRAIN AND THE WORLD

CHAPTER 1
CAN YOU TRUST YOUR BRAIN?

Your brain lies to you a lot. We're sorry to have to break the news to you, but it's true. Even when your brain is doing essential and difficult stuff, you're not aware of most of what's going on.

Your brain doesn't intend to lie to you, of course. For the most part, it's doing a great job, working hard to help you survive and accomplish your goals in a complicated world. Because you often have to react quickly to emergencies and opportunities alike, your brain usually aims to get a half-assed answer in a hurry rather than a perfect answer that takes a while to figure out. Combined with the world's complexity, this means that your brain has to take shortcuts and make a lot of assumptions. Your brain's lies are in your best interest—most of the time—but they also lead to predictable mistakes.

One of our goals is to help you understand the types of shortcuts and hidden assumptions that your brain uses to get you through life. We hope this knowledge will make it easier for you to predict when your brain is a source of reliable information and when it's likely to mislead you.

The problems start right up front, when the brain takes in information from the world through the senses. Even if you are sitting quietly in a room, your brain receives far more information than it can hold on to, or than you need to decide how to act. You may be aware of the detailed pattern of colors in the rug, the photographs on the wall, and the sounds of birds outside. Your brain perceives many other aspects of the scene initially but quickly forgets them. Usually these things really aren't important, so we don't often notice how much information we lose. The brain commits many lies of omission, as it discards most of the information in the world as soon as it is deemed to be unremarkable.

Lawyers know this principle. Eyewitnesses are notoriously unreliable, in part because they imagine—as most of us do—that they see and remember more details than they really can.

Did you know? **Looking at a photograph is harder than playing chess**

You may think that you know what your brain does, but you actually notice only a small fraction of its activity—and what your brain accomplishes behind your back is some of its hardest work. When computer scientists first began trying to write programs to mimic human abilities, they found that it was relatively easy to get computers to follow logic rules and do complex mathematics, but very hard to get them to figure out what they were seeing in a visual image or to move smoothly through the world. Today's best computer chess programs can beat a grand master, at least some of the time, but any normal toddler can kick the butt of the top programs when it comes to making sense of the visual world.

One difficult step, as it turns out, is identifying individual objects in a visual scene. When we look at, say, a dinner table, it seems obvious that the water glass is one object that is in front of another object, like a vase of flowers, but this turns out to be a sophisticated calculation with many possible answers. You only notice this ambiguity occasionally, when you see something briefly enough to misidentify it, like when that rock in the middle of the dark road suddenly turns into the neighbor's cat. The brain sorts out these possibilities based on its previous experience with objects, including having seen the two objects separately and in other combinations. Have you ever taken a picture in which a tree seemed to be growing out of someone's head? When you snapped the photo, you didn't notice the problem because your brain had efficiently separated the objects based on their different distances from your eyes. Later on, the two-dimensional photo didn't contain the same information about distances, so it looked like the two objects were on top of each other.

Lawyers can use this knowledge to discredit witnesses by tempting them to say they saw something that the lawyer can disprove, casting doubt on the rest of the witness's testimony.

In addition to throwing away information, the brain also has to decide whether to take shortcuts, depending on how it values speed against accuracy in a particular situation. Most of the time, your brain favors speed, interpreting events based on rules of thumb that are easy to apply but not always logical. The rest of the time, it uses the slow, careful approach that's appropriate for doing math or solving logic puzzles. Psychologist Daniel Kahneman won the Nobel Prize in Economics for studying these rules of thumb and how they influence real-life behavior. (His longtime collaborator, Amos Tversky, passed away before he could share the honor.)

The take-home message from their research is that logical thinking requires a lot of effort.

Did you know? **Are we in our right minds?**

When people talk about the "right brain" and the "left brain," they're referring to the two sides of the cortex. While there are some real differences in function between them, the distinctions are often misunderstood.

Most people's speech is controlled by the left side of the brain, which is also responsible for mathematics and other forms of logical problem solving. Curiously, it is also the source of many misremembered or confabulated details, and it is the home of the "interpreter" discussed on the next page. All in all, the left side of the brain seems to have an intense need for logic and order—so intense that if something doesn't make sense, it usually responds by inventing some plausible explanation.

The right side is much more literal and truthful when it reports what happened. It controls spatial perception and the analysis of objects by touch, and excels at visual-motor tasks. Rather than being "artistic" or "emotional," the right brain is simply more grounded. It's a Joe Friday type, and if it could talk, it would probably say, "Just the facts, ma'am."

For example, try to solve the following problem quickly: A bat and a ball together cost $1.10. The bat costs $1 more than the ball. How much does the ball cost? Most people say 10¢, which is intuitive but wrong. (The bat costs $1.05, and the ball costs 5¢.) Mental shortcuts like this are very common: in fact, people are likely to use them in almost all situations unless they're strongly cued that they should be using logic instead. Most of the time, the intuitive answer is good enough to get by, even when it is wrong.

In everyday life, we're not typically asked to solve logic problems, but we are often asked to make judgments about people we don't know very well. Kahneman and Tversky used another approach to show that these judgments aren't logical either. For example, they would start an experiment by telling people about Linda: "Linda is thirty-one years old, single, outspoken, and very bright. She majored in philosophy. As a student, she was deeply concerned with issues of discrimination and social justice and also participated in antinuclear demonstrations." Then they asked people to pick the phrase that seemed most likely to describe Linda from a carefully contrived list of traits.

Most people thought it was more probable that (a) "Linda is a bank teller who is active in the feminist movement" than (b) "Linda is a bank teller." Choice (a) makes intuitive sense because many of Linda's other characteristics—concern about social justice and so on—suggest that she might be active in the feminist movement. Yet that is not the right answer, because everyone who is (a) "a bank teller who is active in the feminist movement" is also (b) "a bank teller." And of course the group in (b) includes other bank tellers who are reactionary or apathetic.

In such a case, even sophisticated participants like graduate students in statistics make the error of reaching a conclusion that directly contradicts logic. This strong tendency to attribute groups of related characteristics to people without much evidence is a quick way of estimating likely outcomes, but it may also be a root cause of many of the stereotypes and prejudices that are common in society.

To make matters worse, many of the stories we tell ourselves don't even reflect what's actually happening in our own heads. A famous study of brain-damaged patients demonstrates this idea. The patients had been treated for severe epilepsy by a surgical operation that disconnected the right and left halves of their brain's cortex, to prevent seizures from spreading from one side to the other. This meant that the left half literally didn't know what the right half was doing, and vice versa.

In one experiment, the scientists showed a picture of a chicken claw to the left side of a patient's brain, where the language areas are located, and a picture of a snow scene to the right side of the brain, which cannot produce speech. Asked to pick a related image from another set of pictures, he correctly chose a shovel with his left hand (controlled by the right side of the brain) and a chicken with his right hand (controlled by the left side of the brain). When asked to explain his choices, he responded: "Oh, that's simple. The chicken claw goes with the chicken, and you need a shovel to clean out the chicken shed." The scientists concluded that the left side of the brain contains an "interpreter" whose job is to make sense of the world, even when it doesn't understand what's really happening.

These problems of throwing away information, taking mental shortcuts, and inventing plausible stories come together in what psychologists call "change blindness." For an example, look at the two photographs. What is the difference between them? (Hint: men of a certain age beware!)

When people look at complicated pictures like the ones shown here, they can identify differences if the images remain still. But if the image flickers during the transition from one to another, then they have a lot more trouble. This happens because our visual memory isn't very good.

Experiments of this sort led psychologists to push their luck and try more outrageous ways of getting people to fail to notice things. In one of our favorites, a researcher approaches someone on the street and asks for directions. While the person is replying, workmen carry a large door between the two people, blocking their view of each other. Behind the cover of the door, the person who asked for directions is replaced by another researcher, who carries on the conversation as if nothing has happened. Even when the second person looks very different from the first, the person giving the directions has only about a 50 percent chance of noticing the change.

In another experiment, subjects watch a video in which three students in white shirts pass a basketball around, while another three students in black shirts pass a second basketball. The viewers are asked to count the number of passes made by the white-shirted team. As the two groups mingle, a person in a gorilla suit walks into the game from one side and walks out the other side, after pausing to face the camera and beat his chest. About half of the viewers fail to notice this event. These experiments illustrate that you perceive only a fraction of what's going on in the world.

We've established that your memory of the past is unreliable and your perception of the present is highly selective. At this point, you probably won't be surprised to hear that your ability to imagine the future also is worthy of suspicion. As Daniel Gilbert explains in *Stumbling on Happiness*,

Myth: We use only 10 percent of our brains

Ask a group of randomly chosen people what they know about the brain, and the most common response is likely to be that we only use 10 percent of its capacity. This belief causes neuroscientists around the world to cringe. The ten percent myth was established in the U.S. more than a century ago, and it is now believed by half the population in countries as far away as Brazil.

To scientists who study the brain, though, the idea doesn't make any sense at all; the brain is a very efficient device, and pretty much all of it appears to be necessary. To stick around so long, the myth must be saying something that we really want to hear. Its impressive persistence may depend on its optimistic message. *If we only use 10 percent of our brains normally, think what we could do if we could use even a tiny bit of that other*

90 percent! That's certainly an attractive idea, and it's also sort of democratic. After all, if everyone has so much spare brain capacity, there aren't any dumb people, only a bunch of potential Einsteins who haven't learned to use enough of their brains.

This brand of optimism has been exploited by self-help gurus to sell an unending series of programs to improve brainpower. Dale Carnegie used the idea to win book sales and influence readers in the 1940s. He gave the myth a big boost by attributing the idea to a founder of modern psychology, William James. But no one has found the 10 percent number in James's writings or speeches. James did tell his popular audiences that people have more mental resources than they use. Perhaps some enterprising listener made the idea sound more scientific by specifying a percentage.

This idea is particularly popular among people who are interested in extrasensory perception (ESP) and other psychic phenomena. Believers often use the ten percent myth to explain the existence of these abilities. Grounding a belief that is outside the realm of science in a scientific fact is nothing new, but it's particularly egregious when even the "scientific fact" is known to be false.

In reality, you use your whole brain every day. If big chunks of brain were never used, damaging them would not cause noticeable problems. This is emphatically not the case! Functional imaging methods that allow the measurement of brain activity also show that simple tasks are sufficient to produce activity throughout the entire brain.

One possible explanation for how the ten percent myth got started is that the functions of certain brain regions are complicated enough that the effects of damage are subtle. For instance, people with damage to the cerebral cortex's frontal lobes can often still perform most of the normal actions of everyday life, but they don't select correct behaviors in context. For instance, such a patient might stand up in the middle of an important business meeting and walk out in search of lunch. Needless to say, patients like this have a hard time getting around in the world.

Early neuroscientists may have had some trouble figuring out the functions of frontal brain areas partly because they were working with laboratory mice. In the laboratory, mice have a pretty simple life. They have to be able to see their food and water, walk over to it, and consume it. Beyond that, they don't have to do much of anything to survive. None of that requires the frontal areas of the brain, and some early neuroscientists developed the idea that maybe these areas didn't do anything much. Later, more sophisticated tests disproved that view, but the myth had already taken hold.

when we try to project ourselves into the future, our brains tend to fill in many details, which may be unrealistic, and leave out many others, which may be important. Depending on our imagined reality as though it were a movie of the future, we are prone to overlook pitfalls and opportunities alike as we plan our lives.

By now, you may be wondering if you can trust anything your brain tells you, but millions of years of evolution lie behind its seemingly peculiar choices. Your brain selectively processes details in the world that have historically been most relevant to survival—paying particular attention to events that are unexpected. As we've seen, your brain rarely tells you the truth, but most of the time it tells you what you need to know anyway.

CHAPTER 2

GRAY MATTER AND THE SILVER SCREEN: POPULAR METAPHORS OF HOW THE BRAIN WORKS

If you want to see what happens when the brain goes out of whack, please don't go to the movies. Movie characters are continually getting themselves into neurological scrapes, losing their memories, changing personalities, and getting schizophrenia or Parkinson's disease (not to mention sociopathy and other psychiatric disorders). The brain goes haywire in Hollywood far more often than in real life, and sometimes it can be hard to tell science from science fiction. Movie depictions of mental disorders span the spectrum from mostly accurate to totally wrong. At their worst, movie depictions of neurological illness can reinforce common, but wrong, ideas about how the brain works.

By far the most common mental disorder in the movies is amnesia. Memory loss in the movies constitutes its own genre, as predictable as boy meets girl, boy loses girl, and boy gets girl back. But instead of losing a love interest, the thing lost might instead be, to pick an example, the awareness that one is a trained assassin, as in *The Bourne Identity* (2002) or *Total Recall* (1990).

Neuropsychologist Sallie Baxendale conducted an extensive survey of memory loss in the movies, going all the way back to the silent era. She sorted incidents into categories, most of which are filled with wrong science but all of which are entertaining. A common dramatic theme is a trauma that triggers memory loss, typically followed by a new start of some kind. Our hero or heroine then has a series of adventures and misadventures, but is able to live

Did you know? Depictions of brain disorders in the movies

ACCURATE	INACCURATE
Memento	*Total Recall*
Sé Quién Eres	*50 First Dates*
Finding Nemo	*Men in Black*
A Beautiful Mind	*The Long Kiss Goodnight*
Awakenings	*Nit-Witty Kitty* (Tom and Jerry)
	Murder by Night
	Les Dimanches de Ville d'Avray

normally and form new memories. Another common cause of amnesia in the movies is a psychologically traumatic event. These events, which satisfy the dramatic need to drive the plot, include anything from killing someone to getting married. As a final twist, a character might regain his or her memory by getting whacked in the head a second time, or through a brilliant act of neurosurgery, hypnosis, or the sight of a significant and well-loved object from the past. Roll credits.

There also seems to be an inverse correlation between the incidence of amnesia and the artistic merit of a television program. Soap operas and situation comedies are rife with such cases. The 1960s television series *Gilligan's Island*, which is loved for its entertainment value rather than its accuracy, over three seasons featured no fewer than three cases of amnesia. Another offender is the movie *50 First Dates* (2004), which portrays a pattern of memory loss that never occurs in any known neurological condition. Drew Barrymore plays a character who collects new memories each day and then discards them all overnight, clearing the way for a brand-new beginning the next day. In this way she is able to tolerate more than one date with Adam Sandler. This pattern—the ability to store memories but subsequently lose them on a selective, timed basis—exists only in the imaginations of scriptwriters who get their knowledge of the brain from other scriptwriters.

The head-bonk model of memory loss can even be traced to precinema literature. Edgar Rice Burroughs, creator of the Tarzan novels, was particularly fond of the concept and applied it to quite a few of his potboiler plots. In one of Burroughs's finer literary moments, *Tarzan and the Jewels of Opar* (1918), he manages to separate memory loss neatly from any other neurological damage:

> His eyes opened upon the utter darkness of the room. He raised his hand to his head and brought it away sticky with clotted blood. He sniffed at his fingers, as a wild beast might sniff at the life-blood upon a wounded paw . . . No sound reached to the buried depths of his sepulcher. He staggered to his feet, and groped his way about among the tiers of ingots. What was he? Where was he? His head ached; but otherwise he felt no ill effects from the blow that had felled him. The accident he did not recall, nor did he recall aught of what had led up to it.

Burroughs may have drawn upon an existing belief that head injury could lead to amnesia. In the 1901 book *The Right of Way*, by Gilbert Parker, a snobbish, drink-sodden attorney named Charley Steele, with a nagging wife and a lazy thief of a brother-in-law, suffers amnesia in a barroom assault. This memory loss allows him to escape his many problems and start life over. He finds a new love and is happy until his memory—and old obligations—return. Hollywood loved this plot, making Charley Steele movies in 1915, 1920, and 1931.

Before 1901, the trail of the idea starts to get cold. What enterprising writer first put to

Did you know? **Head injury and personality**

Head injury can sometimes lead to personality change. In real life, this can occur with blows to the front of the head, which can affect the prefrontal cortex. Typical outcomes include the loss of inhibition and judgment. What is not typical is wholesale transposition of personality. In one episode of *Gilligan's Island*, the girlish Mary Ann develops the delusion that she is the sultry starlet Ginger after a blow to the head. Such delusional behavior might accompany schizophrenia or bipolar disorder—but even then only rarely.

A slightly less implausible scenario occurs in the charming *Desperately Seeking Susan* (1985), in which Rosanna Arquette plays a bored housewife who loses her memory and experiences severe confusion. Although the selective loss of identity after a head injury is implausible, one aspect of what happens next contains a grain of truth. A personal ad and a found article of clothing help Arquette invent a story about her lost identity. She goes on to assume the life and obligations of an adventuress on the lam, played by Madonna. Victims of memory loss will often fill in lost information by creating plausible memories, an act of confabulation that creates the illusion of normal, continuous memory.

paper the thought of a head blow leading to amnesia? The notion does represent an advance: it acknowledges the brain as the seat of thought. After all, Shakespeare presented acts of magic as agents of mental change. Think of Titania in *A Midsummer Night's Dream*, who is induced by the prankish Puck's magic love drops to fall in love with Bottom, who has the head of a donkey.

Perhaps we have unfairly made fun of these depictions of memory loss. After all, psychiatric disorders show more diverse symptoms than the strictly neurological disorders stemming from physical injury or disease. For instance, a psychiatric patient might show selective amnesia in very specialized ways. Also, transient memory loss is known to occur spontaneously, possibly because of miniature strokelike events (see chapter 29). But Hollywood usually tells us that the memory loss starts with an injury or traumatic event, and in this regard the targets of our criticism are fair game. Cinema may be ripe for scientific criticism, but it does provide insight into how people think the brain works.

A conceptual underpinning to many cinematic misconceptions is an idea we will call "brains are like old televisions." Consider a common dramatic convention: after a blow to the head induces memory loss, memory can be restored by a second blow to the head. The existence of this myth points to unspoken assumptions we make about how the brain works. For the second-blow hypothesis to be true, damage to the brain would have to be reversible. Since the likeliest cause of amnesia from a head injury would be a fluid accumulation that pushes on the brain, a therapeutic benefit from a second injury would be pretty unlikely, to say the least.

A likely source of the second-blow idea is our everyday experience with electronic devices, especially old ones. It's well known that hitting an old television in just the right way can sometimes get it to work again. These old devices usually have loose or dirty electrical contacts, suggesting that a properly aimed blow might help reseat a connection and thus restore function. The basic problem here is that brains do not have loose connections as such; synapses join neurons together so tightly that no blow, short of a totally destructive injury, would ever "loosen" them.

Many moviemakers seem to think that brains are understood and organized well enough that neurosurgery is useful as a means of repairing memory loss. It is true that neurosurgery can reduce immediately life-threatening conditions, such as the accumulation of fluid or a tumor that compresses the brain. These conditions would usually be accompanied by severe confusion (as in a concussion) or loss of consciousness. Such a surgery needs to be performed immediately after the problem occurs, presenting screenwriters with the dilemma that the dramatic value of any amnesia would have to be compressed into the trip from the injury site to the hospital. Otherwise neurosurgery is more likely to be an accidental cause of memory loss than a cure for it.

In a more realistic (but totally revolting) depiction of brain injury, we have the sequel to *The Silence of the Lambs* (1991), *Hannibal* (2001), in which gradual invasion (oh, let's not mince words—the cutting up and cooking of a person's brain) causes progressive loss of function.

Did you know? **Can memories be erased?**

In *Eternal Sunshine of the Spotless Mind* (2004), the main character seeks to obliterate memories of a relationship gone wrong by going to a professional outfit that provides such a service for a price. In the movie, the character is strapped down and goes to sleep while technicians rummage through his head. They play back memories and pick out the ones that need to be erased.

One idea implicit in this sequence is that neural activity somehow encodes explicit, movielike representations of remembered experiences. Perhaps the logic is not entirely cracked—experience does appear to be reduced and compressed as it is converted to something that the brain can store—but the result is not a full replay of the event (see chapter 1). Recollection of a visual scene does trigger brain responses that resemble in some ways the responses that arise from viewing a scene for the first time. Another part is less fantastic than it may sound: the idea that one can locate an offending memory, play it back, then erase it like an unwanted computer file. Research in the past few years suggests that recollection of a memory also reinforces the memory. There is good evidence that we "erase" and "rewrite" our memories every time we recall them, suggesting that if it were ever possible to erase specific content, playing it back first might be an essential component.

Putting aside the difficulty of carrying out such brain surgery without killing the patient, here at least we have a situation in which damage to the brain leads to proportional loss of function.

In the thicket of misleading and silly depictions of the brain in popular entertainment, a few counterexamples stand out in which the science is accurate. Scientific accuracy is not necessary for a satisfying dramatic experience, of course, but it does seem possible to maintain accuracy, attract critical approbation, and experience commercial success all at once. Various brain disorders are depicted both accurately and sympathetically in the movies *Memento*, *Sé Quién Eres*, *Finding Nemo*, and *A Beautiful Mind.*

Memento (2000) accurately describes the problems faced by Leonard, who has severe anterograde amnesia. Due to a head injury, Leonard cannot form lasting new memories. In addition, he has difficulty retaining information held in immediate memory and, when distracted, loses track of his train of thought. The effect is cleverly induced in the viewer's mind by showing the sequence of events in reverse order, starting with the death of a character, and ending with a scene that reveals the meaning of all the subsequent events.

The symptoms suffered by Leonard are similar to those experienced by people with damage to the hippocampus and related structures. The hippocampus is a horn-shaped structure that in humans is about the size and shape of a fat man's curled pinkie finger; we have one hippocampus on each side of our brains. The hippocampus and the parts of the brain that link to it, such as the temporal lobe of the cerebral cortex, are needed for the short-term storage of new facts and experiences. These structures also seem to be important for eventual long-term storage of memories; patients with temporal lobe or hippocampal damage, such as from a stroke, often are unable to recall events in the weeks and months before the damage.

In *Memento*, the accident that triggers Leonard's amnesia is depicted with remarkable fidelity, right down to the part of his head that receives an injury, the temporal lobe of the cortex. The resulting loss of function is also accurate, with the possible exception that unlike many patients with similar damage, he is aware of his problem and can describe it. The most famous patient with hippocampal and temporal lobe damage, known only as HM, is not so lucky (or perhaps he is luckier). Since he had an experimental surgery to prevent epileptic seizures, HM lives in a perpetual now, continually greeting people as if for the first time, even if he has spoken to them countless times before (see chapter 23).

The 2000 Spanish thriller *Sé Quién Eres* (I Know Who You Are) presents the case of Mario, whose memory loss stems from Korsakoff's syndrome, a disorder associated with advanced alcoholism. Mario cannot recall anything that happened to him before 1977, has difficulty forming new memories, and is often confused. Yet his psychiatrist finds herself drawn to him.

In Mario's case, his memory defects result from damage to his thalamus and mammillary bodies, which is caused by thiamine (vitamin B_1) deficiency resulting from the long-term malnutrition that often accompanies severe alcoholism.

A final example of memory loss in the movies comes from the animated feature *Finding Nemo* (2003). The sufferer in this case is not a human being, but a fish. Dory is friendly but has severe difficulty forming new memories. Like Leonard, she loses her train of thought when distracted. We could complain that it is unrealistic to expect much cognitive sophistication from a fish, but considering the egregiousness of the worst cinematic offenses, we will score this as a minor infraction. What is realistic in this movie is the feeling of being lost that Dory experiences as she finds her way through life, and the way that she can be annoying, even (and perhaps especially) to those close to her.

Did you know? **Schizophrenia in the movies—*A Beautiful Mind***

A Beautiful Mind (2001) dramatizes the life of the mathematician John Nash, presenting the experience of descending into schizophrenia in great detail. The Nash character (in a somewhat loose adaptation of the real Nash) experiences hallucinations and starts to imagine causal links between unrelated events. His growing paranoia about the motives of those around him and his inability to critically reject these delusions gradually alienate him from colleagues and loved ones.

These are classic signs of schizophrenia, a disorder that is caused by changes in the brain induced by disease, injury, or genetic predisposition. Schizophrenia typically strikes people in their late teens and early twenties and affects more men than women. As many as one in one hundred people experience symptoms of schizophrenia at some point in their lives. The hallucinations experienced by the Nash character in the movie are visual; the real-life Nash has experienced auditory hallucinations of a similar nature.

While much of the movie is scientifically accurate, one significant error is that Nash is cured by the love of a good woman. Schizophrenia is not a romantic event; it is a physical disorder of the brain. Some degree of recovery is possible: patients may have periods of normal function interspersed with symptomatic periods, and symptoms disappear in as many as one in six schizophrenics. The reasons for remission, however, are currently not known. The error made in the movie is reminiscent of the old myth that schizophrenia is caused by a lack of mother love, an idea that has no support, is refuted by evidence, and makes mothers—and other loved ones—of schizophrenics feel guilty for no good reason.

This brings us to a striking recurring theme in the accurate depiction of memory loss: the sympathetic portrayal of the sufferer. In inaccurate depictions, the victim is often regarded as a figure of fun or even ridicule. However, the plight of accurately portrayed sufferers is almost always rendered poignantly and, in the best cases, captures the feeling of what it is like to have a disorder.

CHAPTER 3

THINKING MEAT: NEURONS AND SYNAPSES

In his short story "They're Made Out of Meat," Terry Bisson describes alien beings with electronic brains who discover a planet, Earth, on which the most sophisticated organisms do their thinking with living tissue. The aliens refer to brains as "thinking meat." (Gross, we know.) The idea that your brain can generate dreams, memory, breathing, and every mental process in your life may seem hard to believe—but it's true.

This is particularly impressive in view of the brain's size. Considering its many functions, the brain is packed into a very small space. Billions of neurons and additional supporting cells communicate with one another using an astronomical number of synaptic connections—and the entire operation fits into an object weighing about three pounds, the size of a small cantaloupe.

Like a cantaloupe—and the rest of your body—your brain is made of cells. Brain cells come in two types: neurons, which talk to one another and to the rest of the body, and glial cells, which provide essential support to keep the whole show going. Your brain is made up of about one hundred billion neurons—which have a long, skinny, complicated shape—and many more glial cells.

From a distance, the brains of different animals do not look alike. (Compare the shrew and whale brains in the picture.) They all work according to the same principles, however.

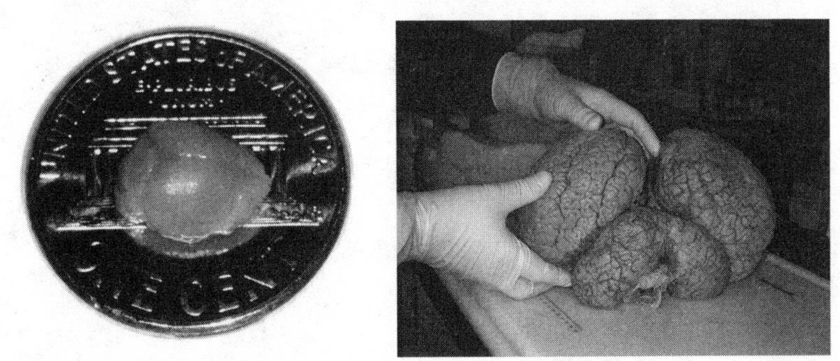

Signals within a neuron are carried by electricity. Each neuron has a net excess density of negative charge on the inside of the membrane that surrounds it relative to the outside, due to an uneven distribution of positive and negative ions like potassium and chloride. This unequal distribution of charge creates a voltage difference across the membrane, like a much smaller version of the voltage difference that allows a nine-volt battery to give a shock to your tongue. (Actively moving ions across the membrane to maintain this charge distribution requires more energy than anything else that the brain does.)

To send electrical signals from one part of the neuron to another, the neuron opens channels that allow the ions to move across the membrane, creating a current that carries an electrical signal down the membrane. Neurons receive inputs through branched, treelike structures called dendrites, which put together information from a bunch of different sources. The neuron then sends an electrical signal down a long, wirelike structure, called an axon, which triggers a chemical signal to another neuron, and so on. Axons can carry signals over long distances; your longest axons run from your spine to the tips of your toes. In contrast, the longest known axons in whales are sixty feet (about twenty meters) in length. The longest axons belonging to the shrew, whose brain is pictured on the penny, are a mere two inches (about five centimeters). In all cases, electrical signals spread using similar molecules and according to the same biological principles.

Let's look at this process in more detail. Neurons pass information down their axons by

Did you know? **Your brain uses less power than your refrigerator light**

Neurons and synapses are so efficient that the brain uses only twelve watts of power—yet it can do a lot more than the little light in the back of your refrigerator. Over the course of a day, your brain uses the amount of energy contained in two large bananas. Curiously, even though the brain is very efficient compared to mechanical systems, in biological terms, it's an energy hog. The brain is only 3 percent of the body's weight, but it consumes one-sixth (17 percent) of the body's total energy. Unfortunately, that doesn't mean that you should snack more to keep your energy up when you're studying. Most of the brain's energy costs go into maintenance, keeping you ready to think by maintaining the electric field across each neuron's membrane that allows it to communicate with other neurons. The added cost of thinking hard is barely noticeable. Look at it this way: you're always paying to support your brain, so you might as well use it!

generating small electrical signals that last a thousandth of a second. These signals are called "spikes" because they represent sudden increases in the electrical currents in a neuron (see graph). Spikes—known to brain geeks as action potentials—look the same whether they come from squid, rats, or Uncle Fred, making them a huge success story in the evolutionary history of animals. Racing down axons at speeds up to several hundred feet per second, spikes bring signals from your brain to your hand fast enough to escape the bite of a dog or the heat of a frying pan. They help all animals get away from imminent danger—fast.

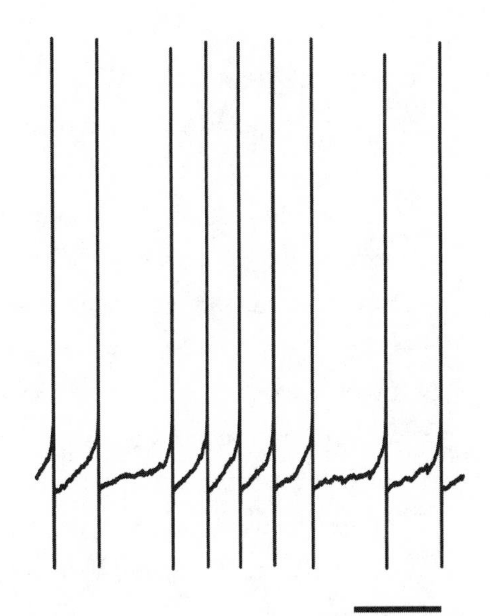

Spikes conclude their business when they arrive at the axon's end. At that point, neurons assume their other identity, as chemical signaling machines. Each neuron in the brain receives chemical signals from some neurons and sends chemical signals to others. Communication between neurons relies on chemicals called neurotransmitters, which are released from small areas at the end of the axon when triggered by the arrival of a spike. Every neuron makes and receives up to several hundred thousand chemical connections, called synapses, with other neurons. Neurotransmitters stick to synaptic receptors on the dendrites or cell bodies of another neuron, triggering further electrical and chemical signals. All these steps, from release to detection, can take place in a thousandth of a second.

Synapses are the essential components of communication in your brain. Your thought patterns, basic abilities and functions, and individuality are determined by how strong these synapses are, how many of them you have, and where they are. Just as connections in computers mostly connect internal components of the computer with one another, neurons mostly use synapses to talk to each other within the brain. Only a small fraction of axons form their synapses outside the brain or spinal cord, sending signals to other organs of the body, including the muscles.

In addition to being fast, synapses are also very small. The dendritic tree of a typical neuron is about two-tenths of a millimeter wide. Yet it receives up to two hundred thousand synaptic inputs from other neurons. Indeed, a cubic millimeter of your brain contains as many as a billion synapses. Individual synapses are so small that they contain barely enough machinery to function and are unreliable, so that arriving spikes often fail to cause any release of neurotransmitter at all.

Did you know? **Loewi's dream of the neurotransmitter**

Back in 1921, it wasn't clear how neurons, or even cells in general, talked with one another. German scientist Otto Loewi made a key observation when he studied how the heart receives signals to speed up or slow down. He was convinced that the vagus nerve, a long nerve that comes from the brainstem and attaches to the heart, secreted a substance to slow the heartbeat. In his laboratory, he carefully dissected the hearts of frogs with the vagus nerve attached. When he stimulated the vagus nerve with electric shocks, the heart slowed down. How did this happen? Loewi's hypothesis was that something came out of the nerve to cause this effect, but he didn't know how to test this idea with an experiment.

Stuck, he did what many people do: he slept on it. One night he woke up, struck with an insight on how to do the experiment. Satisfied, he went back to sleep. The next morning . . . nothing. He couldn't recall what experiment to do. The next time he had the dream, he took care to write down his idea. Unfortunately, the next morning he couldn't read his own writing. Luckily, he had the dream again. This time he didn't wait: he got up, went to the laboratory, and did the experiment that would win him the Nobel Prize in Physiology or Medicine in 1936.

The experiment was a simple one. He placed two frog hearts in two vessels joined by a narrow tube. One heart had the vagus nerve still attached. When he electrically stimulated the heart with the nerve attached, it slowed down. Then, after some delay, the second heart began to slow down as well. This simple experiment demonstrated the existence of what he unpoetically called *Vagusstoff*, a substance (*stoff*) that comes out of the vagus nerve of one frog heart to slow the beat of the other heart. *Vagusstoff*, now called acetylcholine, is one of dozens of neurotransmitters that neurons use to communicate with one another.

It's odd that synapses are small enough to be flaky, but this appears to be a widespread phenomenon. Synapses reach a similar minimum size in the brains of various animals, including mice and people. No one is sure why individual synapses have evolved to be small and unreliable, but one possible reason is that the brain may work better if it's packed with a tremendous number of them. This may be a trade-off that stuffs the most function into the smallest possible space.

For the brain to accomplish its many duties, neurons have to take on very specific tasks. Each neuron responds to a small number of events, such as hearing a particular sound, seeing someone's face, carrying out a certain movement—or other processes that aren't observable from the outside. At any given moment, only a small fraction of your neurons, distributed all over your brain, are active. This fraction is ever shifting; the whole game of thinking depends on which neurons are active and what they are saying to each other and to the world.

Neurons in all animals are organized into local groups that serve the same broad purpose, such as detecting visual motion or planning eye movements. In our own brains, each division can have billions of neurons, with many subdivisions; in a rat, millions, with fewer subdivisions; in a squid or insect, thousands of neurons (though in these tiny creatures' brains, different parts of individual neurons may do multiple things at once). Each of these divisions contains its own distinctive types of neurons, particular patterns of connection, and connections with other brain structures.

Scientists first learned about the functions of different parts of the brain by studying people with brain damage. Sadly, World War I was an especially rich source of data. Soldiers often survived head wounds because high-velocity bullets cauterized their wounds, preventing a fatal loss of blood or even infection. But the soldiers exhibited a baffling range of symptoms, which depended on the location in the brain that was damaged. Modern neurologists still publish papers on patients who have brain damage, most commonly from strokes. Indeed, a few patients with very rare types of damage actually support themselves by participating in paid studies.

Scientists can also figure out what a neuron does by tracking its activity under different conditions, by stimulating it, or by tracing its connections to other brain areas. For example, motor neurons in the spinal cord receive signals from neurons in the cortex that generate basic movement commands. In turn, these spinal cord neurons send signals to the muscles, causing them to contract. If scientists electrically stimulate only the spinal cord neurons, the same muscles contract. Together, these results make it clear that spinal cord motor neurons are responsible for executing movement commands that are generated at higher levels of the brain, although there is still plenty of controversy over exactly what aspect of the movement is specified by these commands.

To learn to get around in your brain, you need a quick tour of its parts and what they do. The **brainstem**, as the name suggests, is at the very bottom of the brain, where it attaches to your **spinal cord**. This region controls basic functions that are critical for life, like reflexive movements of the head and eyes, breathing, heart rate, sleep, arousal, and digestion. This stuff is really important, but you don't usually notice it happening. A bit higher up, the **hypothalamus** also controls basic processes that are important to life, but it gets the fun jobs. Its responsibilities include the release of stress and sex hormones and the regulation of sexual behavior, hunger, thirst, body temperature, and daily sleep cycles.

Emotions, especially fear and anxiety, are the responsibility of the **amygdala**. This almond-shaped brain area, located above each ear, triggers the fight-or-flight response that causes animals to run away from danger or attack its source. The nearby **hippocampus** stores facts and place information and is necessary for long-term memory. The **cerebellum**, a large region at the back of the brain, integrates sensory information to help guide movement.

Sensory information entering the body through the eyes, ears, or skin travels in the form of

Did you know? Is your brain like a computer?

People have always described the brain by comparing it to the latest technologies, whether that meant steam engines, telephone switchboards, or even catapults. Today people tend to talk about brains as if they were a sort of biological computer, with pink mushy "hardware" and life-experience-generated "software." But computers are designed by engineers to run like a factory, in which actions occur according to an overall plan and in a logical order. The brain, on the other hand, works more like a busy Chinese restaurant: it's crowded and chaotic, and people are running around to no apparent purpose, but somehow everything gets done in the end. Computers mostly process information sequentially, while the brain handles multiple channels of information in parallel. Because biological systems developed through natural selection, they have layers of systems that arose for one purpose and then were adopted for another, even though they don't work quite right. An engineer with time to get it right would have started over, but it's easier for evolution to adapt an old system to a new purpose than to come up with an entirely new structure.

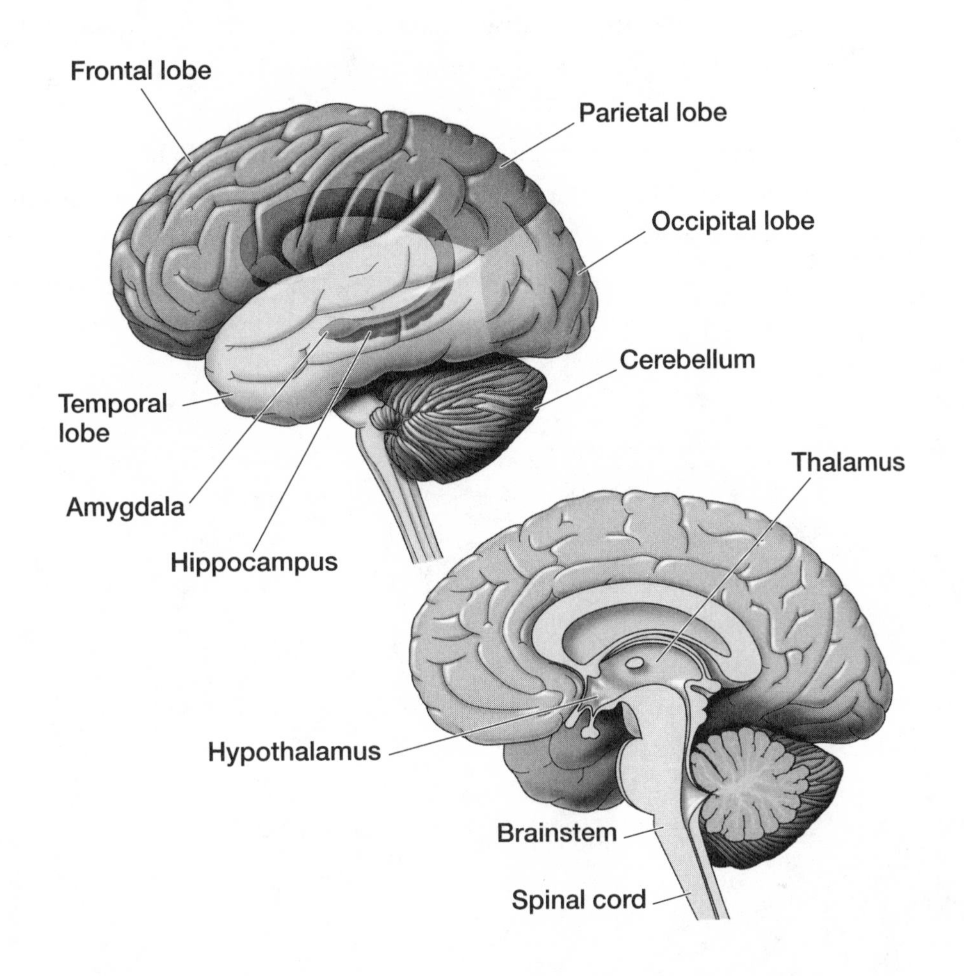

spikes to the **thalamus**, in the center of the brain, which filters the information and passes it along, as more spikes, to the **cortex**. This is the largest part of the human brain, making up a little over three-fourths of its weight, and it is shaped like a large crumpled-up comforter that wraps the top and sides of the brain. The cortex originated when mammals first showed up, about 130 million years ago, and it takes up progressively more and more of the brain in mice, dogs, and people.

Scientists divide the cortex into four parts called lobes. The **occipital lobe**, in the back of your brain, is responsible for visual perception. The **temporal lobe**, just above your ears, is involved in hearing and contains the area that understands speech. It also interacts closely with the amygdala and hippocampus and is important for learning, memory, and emotional responses. The **parietal lobe**, on the top and sides, receives information from the skin senses. It also puts together information from all the senses and figures out where to direct your attention. The **frontal lobe** (you can probably guess where that's located) generates movement commands, contains the area that produces speech, and is responsible for selecting appropriate behavior depending on your goals and your environment.

Together, the combination of these abilities in your brain determines your own individual way of interacting with the world. In the rest of the book, we'll take these abilities in turn and tell you what's known about how the brain accomplishes its everyday tasks.

CHAPTER 4

FASCINATING RHYTHMS: BIOLOGICAL CLOCKS AND JET LAG

Remember when you were a kid and Uncle Larry bet that you couldn't walk and chew gum at the same time? It may have seemed like a lame bet, but when you won his nickel, you were proving yourself to be a remarkably sophisticated animal.

Walking or chewing demonstrates your brain's ability to generate a rhythm. Animals can generate cycles on a wide range of time scales, from seconds (heartbeat, breathing), to days (sleeping), to a month (menstrual cycles), and even longer (hibernation). All these rhythms are generated by built-in mechanisms and adjusted based on external events or commands.

Your ability to generate rhythms simultaneously shows that your brain can generate multiple patterns at once, often independently. Walking involves a tightly coordinated set of events in which your left leg is instructed to rise, move forward, and then lower, as your body simultaneously moves forward. Your right leg follows close behind. The sequence of events has to happen smoothly and in order. These commands are generated mainly by a network of neurons in your spinal cord, all working together as what's called a central pattern generator—central because commands originate here and go to the muscles. This pattern generator can work on its own, since headless cockroaches and chickens can produce walking movements, but they still need their brains to keep everything coordinated and to negotiate obstacles. Chewing is driven by another network of neurons distributed through your brainstem to generate repeated jaw movements. The networks for walking and chewing can work independently (or together, as Uncle Larry discovered).

Practical tip: **Overcoming jet lag**

When you travel, the clocks in your body are able to shift by about an hour per day to reset and get synchronized with the world again. However, you can use your knowledge of circadian rhythms to help you get over jet lag more quickly. The best way to adjust your brain's circadian rhythm is to use light. Melatonin supplements are a distant second. Both are more effective than simply getting up earlier or later and work better than other tricks such as exercise. Here are some guidelines for using light and melatonin to help your body adjust.

• **Get some afternoon light.** The best way to adjust your circadian rhythm is to take a dose of light when your brain can use it as a signal. Light does different things to your circadian rhythm depending on the time of day, just as the timing of your push on a swing affects its movement. In the morning—or, rather, when your body thinks it is morning—light helps you wake up. Exposure to light at this time will get you up earlier the next day—as if the light is telling your body that this time is morning. Exposure to light at night, on the other hand, will get you up later the next day, as if the light is telling your body that the day is not over yet, so it needs to stay awake longer.

So when you fly east, such as from the Americas to Europe or Africa, you should go outside to get some bright light a couple of hours before people back home start to wake up. Finding a source of light is easy at this time because at your destination it is afternoon.

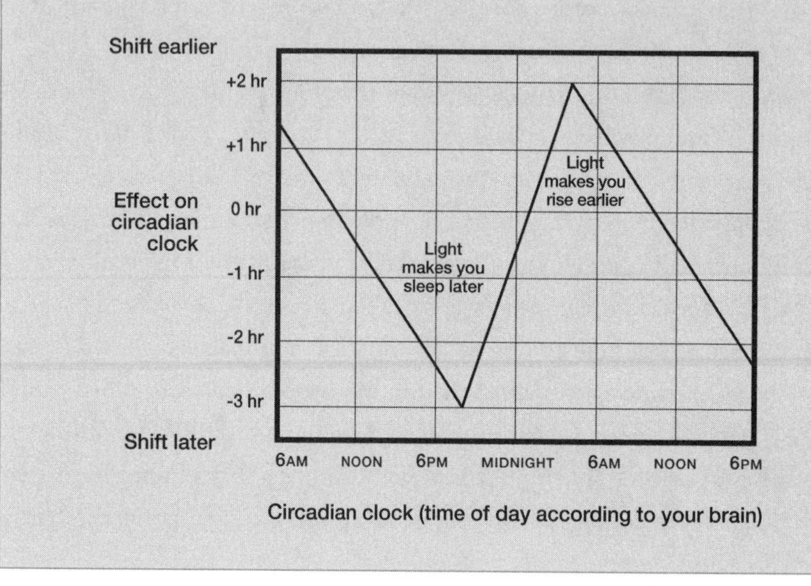

This should help you get up more easily the next day. If you've traveled east across eight time zones or more, try to avoid light first thing in the morning (when it's evening at home), because that will push your clock in the wrong direction. Conversely, when you fly west (from Europe or Africa to the Americas), make sure to get a dose of bright light when you feel sleepy, before it's bedtime back at the place where your flight started.

The simple way to remember both these rules is as follows: On your first day at your destination, get some light in the afternoon. On each subsequent day, as your brain clock adjusts, get some light two or three hours earlier. Lather. Rinse. Repeat.

• **Put out that bedside light!** Enhancing your brain's built-in morning or evening feeling is usually easy because it will still be daytime outside when you need the light. However, it is important to avoid the pitfall of accidentally doing the opposite. Getting light at the wrong time can set your clock in the wrong direction. So if you can't sleep at night, don't turn on the light! Artificial light is less effective than daylight in setting your clock, but you should still avoid it.

• **For long trips, pick a virtual direction.** If you are doing something really crazy like going halfway around the world (Bombay to San Francisco, or New York to Tokyo), decide which way to shift your clock (later each day or earlier each day) and stick with that plan. For most people, but not all, the easiest thing is to pretend you are going west (through Chicago or Honolulu) and get that dose of sun in the very late afternoon. Think of it as a layover for your circadian rhythm.

• **When going east, take melatonin at night.** Light exposure produces melatonin with some time delay, so a pulse of melatonin at night encourages sleep and prepares the next cycle of your clock. As a result, melatonin is elevated in the body clock's evening.

Taking melatonin helps a little if done at the right moment of your circadian rhythm. A dose of melatonin when your body thinks bedtime is soon will help you get up earlier the next day—and help you get to sleep earlier the next night. At your destination, take it at nightfall, or even in the middle of the night. However, for reasons that are not known, melatonin is only helpful if you are going east.

Melatonin's effect is small, shifting your waking time by up to an hour per day. Exercise has a similar effect, and should be done at the same time of day. What we don't know is whether melatonin or exercise does any additional good beyond the benefit of bright light.

You don't always realize it, but you're always falling. With each step you fall forward slightly. And then catch yourself from falling. Over and over, you're falling. And then catching yourself from falling. And this is how you can be walking and falling at the same time.
—Laurie Anderson, *Big Science*

Before we get too impressed with ourselves, we should note one more thing: generating repetitive patterns is a universal feature of animal life. For instance, scientists have studied rhythmic swimming in lampreys, an odd-looking jawless fish that resembles a long thin sock with a ring of teeth at one end. Likewise, they study rhythmic chewing in lobsters, which have relatively simple nervous systems. Lobsters are also interesting because two chewing patterns are directed by a network of only thirty neurons, which adjust themselves and the connections among them throughout life. (And they taste great with melted butter.)

Some patterns are automatic, such as your heartbeat or breathing, but these rhythms can still be controlled. For instance, your heartbeat rhythm, which is generated in your heart itself, can be sped up or slowed down by commands sent by your central nervous system (see chapter 3). Your neuronal network for breathing, which is in your brainstem, can act completely on its own; you don't normally think about breathing. It can also be under close control, as when you hold your breath.

A particularly useful rhythm, found in almost every animal that scientists have studied, is the daily sleep-wake cycle, the circadian rhythm. Circadian rhythms help animals anticipate when light, heat, and food are likely to be available. The circadian rhythm can run on its own, on an approximately twenty-four-hour cycle, and can be reset by correctly timed light exposure. It's synchronized with the daily cycle of light and darkness, which is detected by your eyes. Your circadian rhythms regulate a host of activities, including when you need to sleep, your body temperature, and when you get hungry.

However, these days, circadian rhythms can also trip you up. Nearly anyone who has traveled long distances by airplane has experienced the problem of jet lag. For instance, we wrote part of this book at a study center in Italy. We loved the beautiful setting and the opportunity to get away from our day jobs and concentrate on writing, but there was one catch early in the trip: we found ourselves writing at three in the morning. At breakfast, we engaged in fascinating conversations with other residents, yet sometimes we could barely keep our eyes open.

Jet lag is a product of modern transportation: horseback, dogsled, and even car travel are slow enough that the circadian rhythm can adjust to keep itself in sync with local time. Indeed, the first report of jet lag came in 1931, when two pioneering aviators, Wiley Post and Harold Gatty, flew around the world in a little less than nine days. They experienced the symptoms we recognize today: difficulty getting to sleep, drowsiness, lack of alertness, and digestive problems.

Jet lag happens when your circadian rhythm has different timing from the external day-night cycle in the world. As a result, your brain wants to sleep when it should be awake and vice versa. The brain has a master clock, which normally sets the rhythms for body temperature, hunger, and sleep. With jet lag, these rhythms can get out of sync with each other, causing symptoms like being hungry in the middle of the night.

How light drives circadian rhythms can be explained by an analogy to a child on a swing. The child and swing have a natural period over which one cycle of swinging tends to occur, but if you push the swing, it will change speed. Push when the swing is going forward to make it go sooner; push when it is coming back, and it goes later. In this way, you can alter the starting time of your daily cycle, by exposing yourself to light. To influence your circadian rhythms, though, you must be in the light at the right time of day.

Light acts on circadian rhythms by driving cycles of activity in a tiny region at the bottom of your brain called the suprachiasmatic nucleus, which acts as the master clock. The suprachiasmatic nucleus receives signals from the eye and also generates its own rhythm. Indeed, cells

Practical tip: **Frequent jet lag and brain damage**

Jet lag is not simply annoying; in repeated doses, it can be dangerous to your brain's health. People who frequently cross many time zones can experience brain damage and memory problems. In one study, flight attendants with five years of service who repeatedly took less than five days between long trips were compared to flight attendants who had two weeks or more between trips. (That's still a lot of flying!) Both groups flew the same number of miles overall. The short-interval group had less volume in the temporal lobe—a part of the brain involved in learning and memory. This group also had problems on a memory test, suggesting that frequent travel had damaged their brains.

The brain damage probably resulted from stress hormones, which are released during jet lag and are known to damage the temporal lobe and memory. Luckily, unless you work for an airline, you probably don't need to worry about this problem, since very few people fly across multiple time zones more often than every two weeks. More likely to be at risk are people who do shift work. Like repeated jet travel, frequent drastic changes in working hours are likely to cause stress on the body and brain.

Speculation: Morning people and night people

A tendency to function better at very early or very late hours might result from having a natural circadian cycle that is not exactly twenty-four hours long. A twenty-three-hour period would encourage early rising in people whose bodies are impatient for the day to begin, while the twenty-five-hour person is still whacking away at the snooze alarm.

People with long circadian periods might also make different adjustments to jet lag. On average, more people report difficulties when forced to rise earlier (as in eastward travel) than when forced to rise later (as in westward travel). Difficulties with eastward travel might be associated with periods longer than twenty-four hours. If this is the case, then morning people might have more trouble with westward travel, and night people more trouble with eastward travel—and both these traits would correlate with the natural cycle of a person's clock.

You can help us test these ideas by taking a quiz to see how you score. Report your answers on our Web site at http://welcometoyourbrain.com, and see the results from others.

Quiz

1. When during the day are you most alert? (a) morning or (b) evening or night.
2. In the first two days after a long-distance flight, is it harder for you to adjust to (a) westward travel or (b) eastward travel?

Scoring your answers

Our hypothesis predicts that most people would fall into one of these two categories:

Natural circadian period less than twenty-four hours (morning person type): 1) a, 2) b.

Natural circadian period more than twenty-four hours (night person type): 1) b, 2) a.

from the suprachiasmatic nucleus grown in a culture dish generate patterns of increasing and decreasing activity on an approximately twenty-four-hour cycle. These cells are necessary for normal circadian rhythms; animals with damage to the suprachiasmatic nucleus wake and sleep at odd hours.

Light also triggers the production of the hormone melatonin, which is made by the pineal gland, an organ the size of a large pea that hangs at the bottom of your brain, near the hypothalamus. Melatonin levels start rising in the evening, peak around the onset of sleep, and go down again in the early morning before you wake up.

Incidentally, the pineal gland has quite a romantic history. Several hundred years ago, the philosopher René Descartes thought the pineal gland was the source of consciousness because there was only one of it, and there is only one of you. That was wrong. It just goes to show that even the smartest people can make mistakes when they construct arguments out of thin air.

Most people have a circadian period that is not exactly twenty-four hours, but we don't usually notice because the sun helps keep us on time. When people are left in a room with no lighting cues, they inevitably drift across the clock, and eventually wake, eat, and sleep at hours out of synchronization with the rest of the world.

Blind people, who do not have a way for light information to get from the eye to the brain, natually experience this sort of circadian drift. As a result, the blind often have disrupted sleeping patterns. This shows that physical activity and social cues are not enough to keep people's rhythms in sync. The same is true of blind fish that live in caves; these critters don't seem to ever sleep. The dependence of daily habits on light is indeed universal.

CHAPTER 5

BRING YOUR SWIMSUIT: WEIGHT REGULATION

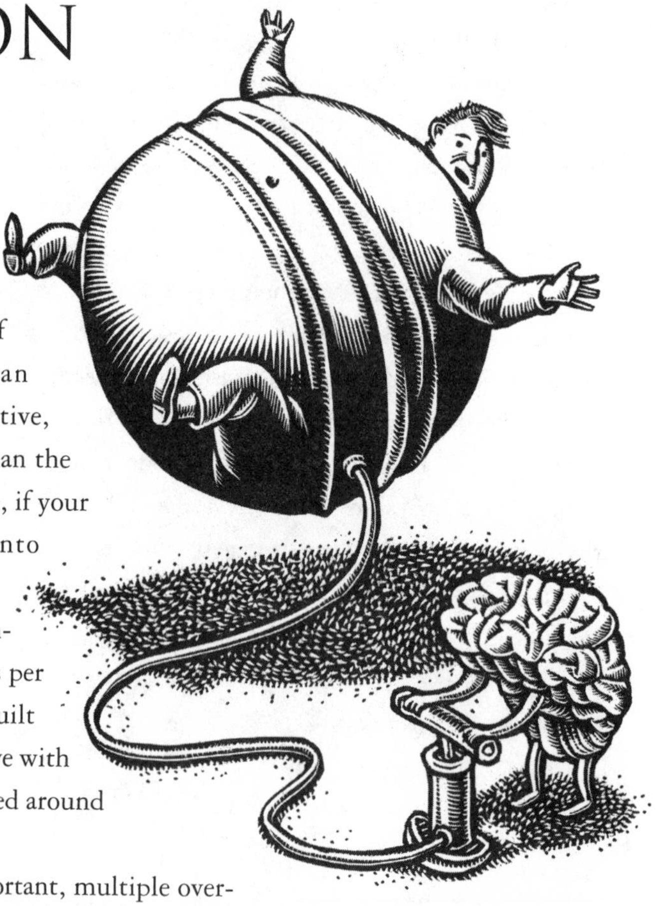

The sad truth is that your brain isn't going to help you if you get fat. From an evolutionary perspective, fat is much better than the alternative, starving to death. Of course, if your brain were smarter, it would take into account that food is abundant in the modern world and that obesity is responsible for three hundred thousand deaths per year in the U.S. But our brains aren't built that way, so we simply have to learn to live with weight regulation systems that developed around the need to store food.

Because weight regulation is so important, multiple overlapping systems work toward keeping your weight at the level that your brain considers appropriate, which is sometimes called your "set point." For example, scientists know of more than a dozen neurotransmitters that tell the body to increase weight, and more than a dozen that tell the body to decrease weight. When you try to change your weight by eating less, your brain falls back on tricks to keep your weight at its preferred level. One is to decrease your resting metabolic rate, which is the amount of energy that you use when sitting still. Another is to make you hungry, so that you'll want to eat more. Finally, your brain may try to fool you in the ways that we discussed in chapter 1. When you find yourself acting as if cake doesn't have as many calories if you eat it in tiny bites from someone else's plate, you're falling for your brain's lies.

Your brain uses several indicators to keep track of your body's energy needs. A hormone called leptin is produced by fat cells and released into the blood. Leptin tells the brain not only how much fat is present in the body but also how fat levels are changing. When your body fat decreases, leptin levels in the blood fall sharply, telling your brain that the body needs more energy. These declining leptin levels trigger hunger and weight gain. In contrast, when leptin levels increase, animals reduce their food intake and lose weight, and people report being less hungry. Leptin receptors in the brain are found in the arcuate nucleus of the hypothalamus, a part of the brain that is an important regulator of many basic systems, including body temperature and sexual behavior. Leptin also acts at other sites in the brain and elsewhere in the body, influencing metabolism and other regulators of fat storage.

My doctor told me to stop having intimate dinners for four—unless there are three other people.
—Orson Welles

Insulin is another important signal that tells your brain how much stored body fat is available. Produced by the pancreas after meals, it is released into the blood to tell a variety of cells to take up glucose from the blood and store the energy. On average, lean animals have lower levels of circulating insulin than fat animals, though insulin varies much more over the course of each day than leptin. Leptin is a good measure of subcutaneous fat, while insulin is related to the amount of visceral fat, which is a more significant risk factor for diabetes, hypertension, cardiovascular disease, and many cancers.

The brain doesn't like to take fat out of storage for everyday energy needs, saving it instead for emergencies. It's a long-term strategy, just as it's better not to dip into your retirement account to buy gas for your car. Thus, neurons in the hypothalamus and the brainstem also monitor available energy sources to control food intake. For example, fatty acids and a hormone called peptide YY seem to act directly on neurons to reduce eating, while the hormone ghrelin is released around mealtimes to increase hunger and eating. These regulatory systems, probably along with others that are yet to be identified, interact to determine whether your brain detects an energy deficit or a surplus at any given time.

Many of these regulators, including leptin, insulin, and other hormones, act in the brain by influencing opposing groups of arcuate neurons. Melanocortin neurons decrease available energy

Did you know? **Calorie restriction and life extension**

In the 1930s, scientists found that rodents kept on a low-calorie diet lived about 50 percent longer than their freely fed counterparts. To varying degrees, the same effect has been found in yeast, worms, flies, fish, dogs, cows, and even monkeys. Calorie restriction reduces cancer, cardiovascular disease, and other age-related problems in rodents and monkeys. It also protects the brains of rodents with experimentally induced Huntington's disease, Alzheimer's disease, Parkinson's disease, or stroke. It's hard to study lifespan extension in humans because our lives are so long already, but there is evidence that calorie restriction has some beneficial effects on human health, like reducing blood pressure and cholesterol.

There's a catch, of course. We're talking about *really* low-calorie diets, which provide about two-thirds of the calories of a normal diet, while still providing required nutrients, such as vitamins and minerals. Many of the same effects can be achieved by a starve-and-binge strategy organized around a normal calorie intake, where you eat nothing one day and then double your calories the next. Most people couldn't stick to such a diet, but there are a few longevity researchers who have been doing it for years.

Calorie restriction seems to work by affecting insulin signaling pathways, which are important regulators of energy storage in the body. Calorie-restricted mice have much lower insulin levels than their well-fed siblings and are much more sensitive to the effects of insulin. Under a normal diet, insulin sensitivity declines with age. This effect is even stronger under a high-calorie diet. Declining insulin sensitivity is a predictor of type 2 diabetes.

Changes triggered by calorie restriction begin with the activation of a receptor for a group of signaling molecules called sirtuins. In mammals, the receptor is called SIRT1, and it is expressed throughout the body. A chemical called resveratrol, which is found in red wine, increases production of SIRT1 in rodents. Resveratrol promotes health and extends the lifespan of mice that are fed a high-calorie diet. The drug doesn't prevent the mice from gaining weight, but it does make them live 15 percent longer. We like red wine too, but don't get your hopes up just yet: the doses used in that study were equivalent to five hundred bottles per day. Another study reported that mice fed resveratrol showed better athletic performance on a treadmill, but those doses were higher still, equivalent to three thousand bottles per day. We couldn't drink that much in a year, let alone in a day. For now, these studies provide hope for your children or grandchildren; at this stage, though, there's not enough evidence for the safety and effectiveness of such supplements to justify their widespread use.

by reducing food intake and increasing energy expenditure. Meanwhile, neuropeptide Y neurons increase available energy by promoting food intake and reducing energy expenditure. Leptin directly activates the melanocortin neurons and inhibits the neuropeptide Y neurons. The process is a bit more complicated than that, though, because the neuropeptide Y (pro-feeding) neurons also strongly inhibit the melanocortin (anti-feeding) neurons. The melanocortin neurons, in contrast, do not have any direct influence on the neuropeptide Y neurons. As a result, this brain circuit is biased toward the promotion of eating and weight gain.

Melanocortin neurons are also found in the brainstem, a part of the brain that regulates fundamental processes like breathing and heart rate. The nucleus of the solitary tract in the brainstem receives input from nerves that originate in the gut, which carry signals related to intestinal expansion or contraction, the chemical contents of the digestive system, and neurotransmitters released in response to nutrients, including some of the ones discussed above. The nucleus of the solitary tract then sends information forward to the hypothalamus, including the arcuate nucleus. Brainstem neurons seem to be particularly important for signaling when an animal is ready to stop eating, through various proteins produced in the gut.

The melanocortin system might seem like a good target for weight-loss drugs, since weight regulation can be strongly affected in mice by genetically altering these receptors and by manipulating the neurotransmitters that activate them. Unfortunately, it may be difficult to avoid side effects because drugs that affect melanocortin receptors also influence blood pressure, heart rate, inflammation, kidney function, and male and female sexual function. Mutations in the melanocortin system in humans are rare and do not account for much of the obesity in the population, though when they occur, they do lead to problems with weight regulation.

When leptin was discovered about ten years ago, researchers were optimistic that it might prove to be the magic bullet that would reduce appetite and cause weight loss. As it turns out, though, many overweight people already have high levels of leptin in their bloodstream but don't respond normally to the hormone, showing what scientists call "leptin resistance." In most people, leptin resistance is a consequence of obesity. This leptin resistance is similar to insulin resistance, which is triggered by weight problems and is the cause of adult-onset diabetes. Obesity caused by overeating causes leptin to become less effective at activating signals that instruct the arcuate nucleus to reduce the body's weight.

Although the discovery of leptin has not led to an effective drug for weight loss, there is a drug based on another pathway that shows some promise. Anyone who's ever gotten the munchies from smoking marijuana knows that pot's active ingredient, delta-9-tetrahydrocannabinol (THC), stimulates hunger even in animals that are well fed. A drug called rimonabant blocks the receptor that responds to THC and reduces food intake even in hungry animals.

Practical tip: Tricking your brain into helping you lose weight

If your brain works against you when you want to lose weight, then how can you achieve the results you want? Basically, you need to arrange your weight-loss strategy to take your brain's reactions into account. Most importantly, that means keeping your metabolic rate as high as possible. It also means finding a strategy that is sustainable. Your brain will always be working toward its own automatically set goals, so any changes you make to your eating and exercise habits will also need to be permanent to remain effective. Temporary changes give temporary results, period. This approach may not sound as glamorous as the latest grapefruit diet, but it does have one substantial advantage: it works.

Your metabolic rate determines how many calories your body burns at rest. Severely low-calorie diets never work in the long run because the very real risk of starvation in our evolutionary past has produced brains that are expert at protecting the body from severe weight loss. One of the main ways that your brain achieves that goal is by slowing down metabolism in times of famine, in some people by up to 45 percent. If your weight was stable on two thousand calories per day, it may also be stable on twelve hundred calories a day after this metabolic compensation kicks in—only now your life is a lot more difficult. Worse yet, when you increase your food intake, you're likely to gain weight before your metabolism adjusts back. Like starvation, sleep deprivation strongly depresses metabolism, so it's important to get enough sleep if you want to keep your weight down. Stress is another culprit, as the stress hormone corticotropin releasing factor tips the body's energy balance in favor of conservation. Metabolism also tends to slow down as you age, which is why people tend to gain weight as they get older, at a rate of about one pound per year.

Exercise is the most effective way to improve this situation, both because the exertion itself triggers your body to increase its use of energy and because muscles burn more calories at rest than fat does. Exercise can boost metabolism by 20 to 30 percent, and the effect lasts up to fifteen hours. Yoga may be a particularly good exercise because many people find that it also reduces stress.

Weight gain and fat storage increase when humans and other animals are fed a few big meals rather than many small ones. Therefore, you should split your calories into small meals spread out over the entire day rather than eating only once or twice a day. In one study, people on a laboratory-controlled diet were able to boost their metabolism by eating in the morning—enough to add two hundred to three hundred calories a day to their diets without gaining weight. This means that a small breakfast pays for itself in metabolic improvement. People who eat the same number of calories gain less weight if they eat in the morning than if they eat in the evening. Of course, it's important to make sure that your frequent meals are actually small! Total calorie intake remains a major determinant of weight, whenever you eat.

A history of repeated weight gain and loss makes it more difficult to maintain a healthy weight. People who've lost at least ten pounds have to eat less (forever) than people who have always been slim. In one study, formerly overweight people had to eat 15 percent fewer calories than their always-thin counterparts to maintain the same weight. For this reason, one of the best gifts you can give your children is to feed them a healthy diet when they're small. Early food exposure influences dietary preferences in adulthood, and eating habits formed in childhood follow many of us around for the rest of our lives.

Contrary to popular belief, eating correctly doesn't involve deprivation and hunger. If you are constantly hungry, you're probably not eating right. Your brain's hunger sensors respond to stomach fullness and to fat and sugar in the bloodstream. To reduce hunger, try combining a large amount of low-calorie food like salad or vegetable soup with a small amount of fat. Finally, find some passion in your life beyond eating. It's much easier to keep your weight down if you have other interesting things to think about. Trips between the television and refrigerator do not count as exercise or as a hobby.

Perhaps more importantly, it has the same effect on those that have already been fed. Animals that eat when they are not hungry may be a fairly good model of human obesity.

In several large clinical trials, obese people who took rimonabant for one year lost about ten pounds more than people who were given a placebo. Treated patients also showed a significant increase in HDL ("good") cholesterol and a decrease in triglycerides, which was partly independent of the weight loss, suggesting that rimonabant has direct effects on lipid metabolism that might reduce heart attack risk. This isn't the kind of weight loss that would change anyone's life, but if it's widely used, the drug is likely to reduce the medical cost of obesity complications. Unfortunately, people in the trial who went off the drug typically gained all the weight back in the following year, so it may need to be taken chronically to maintain weight loss. That's good news for the drug company but bad news for patients.

The receptor that is blocked by rimonabant does not exist to be activated by marijuana, of course, but by brain-synthesized neurotransmitters that are known as endogenous cannabinoids or endocannabinoids. One study reported that people with a mutation in an enzyme that breaks down one of the endocannabinoids, who thus have abnormally high levels of receptor activation, are significantly more likely to be overweight than people without the mutation. This evidence suggests that the cannabinoid system may influence the genetic risk of obesity in the general population. A later study failed to confirm this finding, though, so it's not yet clear whether these mutations are important in many cases of human obesity.

Is the current epidemic of obesity in the U.S. caused by individual differences in genes that help regulate food intake? Not exactly. The efficiency of your cannabinoid and melanocortin systems probably does influence your personal risk of becoming obese, but, in general, people get fat in the modern world because their brains are helping them to store up fat in anticipation of the next big famine. When faced with an excess of good-tasting food, laboratory animals tend to get fat, and so do people. Genetic differences probably determine which people gain weight early in this process and which people require a stronger stimulus, but constant exposure to an excess of tasty food will eventually break down almost anyone's willpower. For this reason, you'd be better off putting your energy into changing your environment so that the available choices are healthy ones rather than spending your mental energy trying to resist the urge to reach for that chocolate bar. Your brain will thank you, and so will your waistline.

PART TWO

COMING TO YOUR SENSES

Looking Out for Yourself: Vision

How to Survive a Cocktail Party: Hearing

Accounting for Taste (and Smell)

Touching All the Bases: Your Skin's Senses

CHAPTER 6

LOOKING OUT FOR YOURSELF: VISION

While skiing downhill one day, Mike May realized he was headed toward a huge dark object too close to dodge. He was sure he was going to die. When he passed through the object, he realized it was a shadow cast by the ski lift.

Such experiences are common in May's life, ever since he had his sight restored by a corneal transplant at age forty-three. May had been blind since a jar of lantern fuel exploded in his face when he was three. However, blindness did not stop him from becoming an excellent skier. He had claimed the world record for speed as a blind downhill skier, following his guide down the mountain at sixty-five miles per hour. During his four decades of blindness, though, his brain had no experience of natural vision. Now, with his vision restored, he has trouble interpreting what he sees. It's especially hard for him to distinguish two-dimensional objects from three-dimensional objects, an essential skill when you are approaching a large two-dimensional shadow.

Your brain interprets many scenes without making you explicitly aware of what's going on. Because May learned to see late in life, the way you might learn a foreign language as an adult, his brain is unable to accomplish many visual tasks correctly, such as figuring out that the large, dark, featureless object in front of him was probably a shadow and not a rock. In general, it's hard for him to figure out which lines or colors are part of one object, and which are part of another object, or even part of the background behind the objects. His case illustrates how difficult and important these processes are in understanding how to see—and how many invisible assumptions your brain needs to make to get the job done.

Vision begins in the eye, which is set up like a camera. A lens in the front of the eye focuses light onto a thin sheet of neurons in the back, called the retina. Retinal neurons are arranged

Did you know? **Animal research and "lazy eye"**

One of the best examples of how animal studies can have unexpected benefits for human medicine comes from research on visual development. Because the two eyes are in different places on the head, they see the world from slightly different angles. This creates a problem for brain development; to create a coherent view, the brain needs to match up the information arriving in the two eyes that comes from the same part of the visual world. It would be hard to specify this matching in advance, since everyone's head is a different size, and the distance between the eyes changes as the body grows. So the brain figures it out by learning to match up information from locations in each eye that are active at the same time, and so presumably are seeing the same place in the visual world. If an animal is deprived of sight in one eye when it's young, then this learning can't happen, and almost all the visual neurons in the brain end up carrying signals from just one eye. If an animal loses sight in one eye at certain young ages (about the first month after birth in cats, longer for people), its brain will learn to interpret information only from the other eye. This pattern can't be reversed later in life. David Hubel and Torsten Wiesel won the Nobel Prize for discovering this process.

A friend of ours has a daughter with strabismus, what people used to call lazy eye, which occurs in 5 percent of children. She has trouble controlling the movement of one eye, leading it to wander off in a different direction from the other one. Twenty years ago, the standard treatment for this problem would have been to keep a patch over the good eye (to train the bad eye to see better). Because of these animal studies, which were undertaken for pure scientific curiosity, we now know that this treatment isn't a good idea, even though it seemed sensible enough at the time. Patching one eye damages brain development because the brain can't learn how to process information from the two eyes together.

You need information from both eyes to judge distances. If you close one eye, and then open that one and close the other, you'll see that the difference between the views is bigger for objects that are closer, and smaller for objects that are very far away. Children who are raised with a patched eye can't compare information from the two eyes, and they have trouble with depth perception as adults. For example, they find it extremely difficult to thread a needle. Because of the animal research, our friend's daughter is being treated with a new training procedure that will let her learn to control her eye muscles without interfering with her ability to see the world in three dimensions later in life.

like a sheet of pixels, each of which detects the intensity of light in a certain region of the visual world. But this causes a problem for the brain, because the retina transforms the three-dimensional world into a pattern of activity in a two-dimensional sheet of neurons, throwing away a lot of the information that's out there. (You may have heard that the retina turns the world upside down, which is true, but it doesn't affect our vision because the brain expects that and interprets the information correctly.)

Three different types of so-called cone cells in the retina detect red, green, or blue colors in bright light; these neurons send increasingly strong signals as the intensity of the light that they detect becomes stronger. Other colors are formed by different levels of activity in combinations of these three cell types. The process is similar to making many colors of paint by mixing primary colors together, but the primary colors are different because light mixes differently from paint. (To see for yourself, put red and green plastic over a couple of flashlights and shine them on the same spot to make yellow light. Mixing red and green paint gives a very different result, brown.) A fourth cell type, called a rod, detects light intensity in dim light but does not contribute to color vision, which is why you can't see colors as well when the lighting is romantic. These rods and cones then communicate with other neurons in the retina, which make additional calculations about the scene. For example, the output cells of the retina carry information about each region's relative brightness compared to nearby areas, not about the absolute brightness of each pixel. This information is then sent into visual areas of the brain, as well as into areas that control movements of the eyes and head.

At each step along the way, neurons are arranged into a map of the visual world, so that information from neighboring points in the scene is represented by the pattern of spikes in neurons neighboring each other in each visual brain area. This is similar to the way that points that are close together in a scene are also close together in a photograph of the scene. Such an organization makes it easier for neurons that represent nearby parts of the visual world to communicate with each other when they're trying to understand their local region of the scene.

The brain must begin by determining the brightness of each part of the object that produced the visual image. You might imagine that this is a simple task—merely a matter of determining how much activity is generated in the neuron that transmits information from that part of the scene. However, this is actually very difficult because neural activity depends on the actual amount of light that reaches the eye, which varies enormously with the characteristics of the object and with the pattern of illumination and shadows in the scene. The same object looks very different in bright sun than under a desk lamp, and different again depending on which part of it is in shadow. The figure on the next page shows that by the time you become aware that you're seeing an image, the brain has already made a bunch of assumptions about the object you're looking at.

In the figure on the left, it's obvious that the square marked A is darker than the square marked B—or is it? The figure on the right shows clearly that those two squares have the same shading. Don't believe us? Cut a piece of paper to cover the extra squares in the left figure and see for yourself.

Have you ever seen a dog moving its head back and forth while staring at something? A lot of animals use this trick to figure out the distance of an object. Closer objects appear to move farther from side to side during this head movement, while more distant objects move less. The brain calculates depth in a scene from many different cues—and a liberal dose of assumptions. For example, depth can be calculated by comparing the views from the two eyes or by determining which objects are in front of other objects. A gravel road going into the distance has two prominent depth cues: the gravel pieces look smaller when they're farther away, and the road edges look closer together. The brain can also use the size of a known object to guess the size of other objects.

Another thing your brain decides automatically is which objects are in a visual image. Mike May has a lot of trouble identifying objects. He can tell the difference between a triangle and a square sitting separately on a table, but he has no idea how many people are in a photograph. The skylights at the mall produce a pattern of alternating bright stripes and shadows across the floor that look, to his brain, exactly like a staircase. After the operation, his wife had to remind him again and again not to stare at women, since he can't get any information from a quick sideways glance the way most men do. He's learned intellectually how to reason through a visual scene and figure out what's in it, to some extent, but this process will never be fast or effortless for him as it is for most of us.

The brain has special ways of recognizing objects of particular importance to us, such as faces. The physical differences between faces aren't all that large—or at least they wouldn't seem that way to a Martian—but we can tell them apart effortlessly. People have tried to devise automated face-recognition systems to identify suspected terrorists in airports and at immigration checkpoints, but their accuracy is terrible compared with human observers. You can see for yourself that your brain treats faces in a special way by looking at the pictures of Margaret Thatcher. The photos at the top look fairly normal to most people—except for being upside down, of course. The bottom pictures are the same images turned right side up, and now you can see that the one on the right is really weird! Both the eyes and the mouth have been turned upside down within the face, but you probably didn't notice that when looking at the top right picture. Of course, which version you prefer may depend on other factors, such as your political orientation.

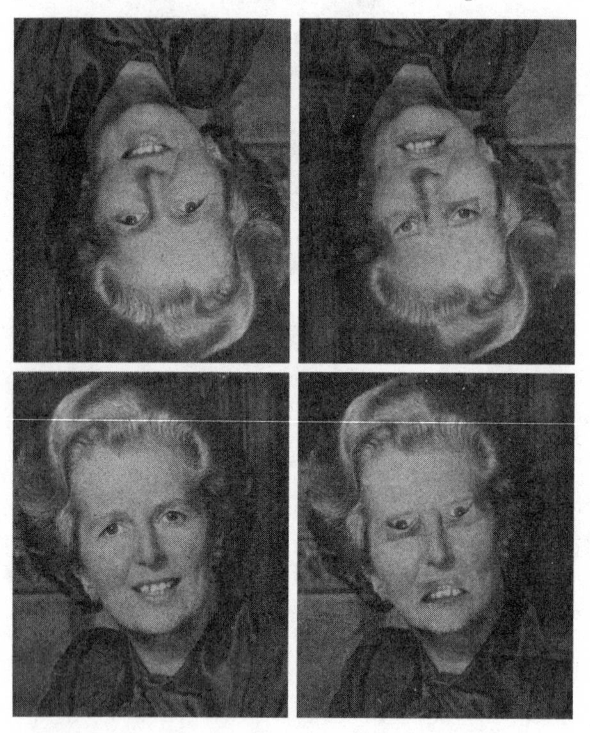

Mike May can't recognize faces at all. He once offered to buy ice cream for a player after Little League practice; only when the puzzled boy politely declined his offer did Mike realize that the player was not his son. Some people who are otherwise normal have the same problem, usually resulting from damage to a brain region called the fusiform face area, which is responsible for the specialized processing of faces. These people can see most objects just fine, but they have a lot of trouble telling people apart, even people they've lived with for years. After a while, most of them learn to memorize what their friends, spouses, or children are wearing when they leave the house so they can recognize them later in a group of people. In May's case, his fusiform face area has not had a chance to develop as it does in people who have grown up with vision.

Right after he had his sight restored, Mike May had to ski with his eyes closed. His brain's motion-detecting cells are as sensitive as a normal person's, but that's a mixed blessing for him. Skiing down the mountain was no longer exhilarating but became frightening as he watched the world zoom past him. For the first time in his life, he became uncomfortable with his wife's driving because he found the sensation of other cars zipping past on the road overwhelming.

Did you know? **The neuron that loved Michael Jordan**

What does it mean to be a fan of a celebrity? One study suggests that it literally means giving space in your brain to that person. There's an old idea that activity in one or a few neurons might signal the identification of a certain object or person, but most neuroscientists don't believe that the brain does it that way. This is because there just aren't enough neurons to account for everything that we can recognize—and because people don't have strokes that eliminate their ability to recognize some people but not others (though some patients lose their ability to recognize people in general, as discussed earlier in this chapter).

In this study, scientists recorded single neurons from the brains of eight people with intractable epilepsy. Surgeons implanted electrodes in the temporal lobe of each patient's brain to help identify the origin of the seizures, and the scientists used these electrodes to record from neurons while the patients were looking at pictures. Some neurons responded specifically to images associated with a particular celebrity (usually an actor, politician, or professional athlete). For example, one neuron fired spikes in response to all photos of Jennifer Aniston—except the one where she appeared with Brad Pitt—and did not respond to pictures of anyone else. Another neuron was activated by photos and drawings of Halle Berry, and even by her printed name. Although this neuron responded to a picture of Halle Berry dressed in her Catwoman costume, it did not respond to the photo of another woman in a Catwoman costume. Other neurons responded to Julia Roberts, Kobe Bryant, Michael Jordan, Bill Clinton, or even famous buildings like the Sydney Opera House. No one is sure what these neurons actually do, though one brain region where they're found is involved in the formation of new memories.

No one is sure why the brain's motion system is so robust that it can function after forty years of blindness, but it might be because motion detection is so important for survival. Whether you're a hungry wolf or a terrified rabbit, there's nothing better than motion for finding the other living things in your visual world.

The brain areas that analyze motion are separate from the ones that analyze shape. In fact, they're in a different part of the brain. The basic motion area detects object movement in a straight line, while higher areas detect more complicated patterns, including expansion (like rain seen through the windshield of a moving car or the opening sequence of *Star Trek*) and spiral motion (like the water swirling down your bathtub drain). These signals are probably important for navigation, as your retina experiences these sorts of motions as you move through the world.

Damage to these brain regions causes motion blindness. People with this disorder see the world as if they were under a strobe light at a disco: first a person is here; then suddenly he's somewhere else. As you can imagine, it's very dangerous to live in a world where it seems like all the other people and objects are capable of random teleportation, so these people have a lot of trouble getting around.

So far we've talked as if our eyes were taking in a continuous scene, something like a movie playing on the retina, which is certainly what it feels like. This is because the brain has ways of smoothing over the world to make your experience feel continuous even when it isn't. However, by now you've probably guessed what comes next: your brain is lying to you again.

Myth: **Blind people have better hearing**

People have long attributed special powers—even magical powers—to blind people. One common idea is that the blind have extra-sharp hearing. However, when tested, blind people are no better at detecting faint sounds than sighted people.

But one old belief about blind people's special abilities is correct. In ancient times, before the invention of writing, the blind were known for their accurate memories of biblical interpretations, which were passed down from one generation to the next as oral traditions. Indeed, blind people do have better memory, especially for language. Since they can't rely on vision to tell them things like "Did I set that glass down on the counter?" they have to use their memory constantly (or else knock a lot of drinks to the floor). Presumably, constant practice helps them sharpen their spatial memory. They also do better than sighted people at other language tasks, including understanding the meaning of sentences. In addition, blind people are better at localizing sounds, which may be another way of keeping track of where things are.

Blind people seem to improve these abilities by taking advantage of brain space that isn't being used for vision. In blind people, verbal memory tasks activate the primary visual cortex, which is involved only in vision in sighted people. Researchers can temporarily turn off a region of the cortex by applying magnetic stimulation to the outside of the skull to interfere with the brain's electrical activity. This interference impairs blind people's ability to generate verbs, which is one of the language tasks that they do especially well, but it has no effect on this task in sighted people (though it does, of course, interfere with their ability to see).

All the time you're awake, your eyes are jumping around the visual world in abrupt movements called saccades, which occur three to five times per second. You can see these movements by watching a friend's eyes. Each eye movement gives the retina a "snapshot" of some part of the visual scene, but the brain must put these still pictures back together to create the illusion of a continuous world. Even neuroscientists don't have much of an idea about how this complicated process works.

To see what is in front of one's nose needs a constant struggle.
—George Orwell

Mike May's experiences illustrate that although vision appears to be one sense, it is really composed of many functions. To most of us, these functions are woven together to form a seamless whole, thanks to a lifetime of development and experience. May's brain has not learned how to lie, or even to tell the truth, fluently. As a result, he can navigate visually 90 percent of the time. That's not as useful as it sounds, though, since he never knows which 10 percent of his perceptions are wrong. Now that he has vision, he's discovered that he can't always trust it. Four years after his sight was restored, Mike May finally figured out how to deal with these problems: he got his first seeing-eye dog since his operation.

CHAPTER 7

How to Survive a Cocktail Party: Hearing

We often think of vision as our most important sense, but perhaps equally essential is hearing. For obvious reasons, deafness makes it difficult to communicate with other people. Deaf people have risen to this challenge by creating their own unique form of language, which uses the hands and eyes instead of the mouth and ears. The barriers to communication between deaf and hearing people are so profound that distinctive deaf cultures have arisen. (For example, in the movie *Children of a Lesser God*, when a deaf woman falls in love with a hearing teacher at the school where she works, the conflict with her loyalty to deaf society threatens their relationship.) How your brain identifies complex sounds like speech is still something of a mystery, although scientists understand quite a bit about how we detect and locate auditory signals.

Whether we're listening to music, birdsong, or the chatter of a cocktail party, hearing begins with a set of pressure waves in the air that we call sound. If we could see the waves caused by a pure tone (a flute note would be the closest everyday example) as they moved through the air, they would look like the ripples you produce when you throw a rock into a pond. The density of the ripples (called frequency) determines the pitch of the tone—shorter distances between waves make high sounds, longer ones make low sounds—and their height determines sound intensity. More complicated sounds, like speech, contain multiple frequencies with different intensities mixed together.

The outer ear transmits these sound waves to an organ in the inner ear called the cochlea (Latin for *snail* because it's shaped like one, as you can see in the drawing). The cochlea contains the ear's sound-sensing cells, which are arranged in rows along a long, coiled membrane. Sound pressure moves the fluid in the ear, causing the membrane to vibrate in different ways depending on the sound's frequencies. This vibration activates the sensors, called hair cells because they have a bundle of fine fibers that stick up from the top of the cell like a punk hairdo. Movement of these fibers transforms the vibration signal into an electrical signal that can be understood by other neurons. Hair cells can sense movement the size of an atom and respond very rapidly (more than twenty thousand times per second).

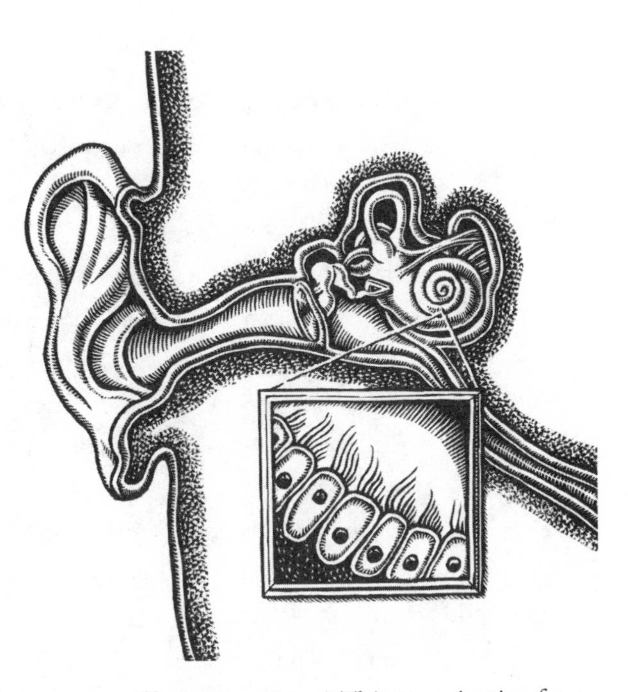

Hair cells at the base of the cochlear membrane sense the highest frequencies. As you move around the coil toward the other end, hair cells become sensitive to lower and lower frequencies. (Imagine the sequence of keys on a piano.) This organization forms a map of sound frequency, which is maintained in many of the brain areas that respond to sound.

Sound information from the two ears is brought together in the neurons of the brainstem. Doctors use this knowledge to help diagnose the causes of hearing loss, based on whether it occurs in one ear or in both. Because neurons within the brain get sound information from both ears, any damage to parts of the brain that process sound causes hearing problems in both ears. For this reason, if you have difficulty hearing in only one ear, the problem is likely to be damage to the ear itself or to the auditory nerve. Hearing loss can also be caused by mechanical problems that interfere with the transmission of sounds from the outside of the ear to the cochlea. This type of hearing loss can be treated with a hearing aid, which amplifies sounds entering the ear. Hearing loss caused by damage to hair cells can only be helped by a cochlear implant (see *Practical tip: Improving hearing with artificial ears*).

The brain has two major goals for sound information: to locate a sound in space, so you can look toward the sound's source, and to identify the sound. Neither of these tasks is easy, and each is accomplished in different parts of the brain. Therefore, some brain-damaged patients have difficulty locating sounds but not identifying them, and vice versa.

Differences in the timing and intensity of sounds reaching your right and left ears help your brain to figure out where a given sound came from. Sounds coming from straight ahead of you (or straight behind you) arrive at your left and right ears at exactly the same time. Sounds coming from your right reach your right ear before they reach your left ear, and so on. Similarly, sounds (at

Practical tip: **How to prevent hearing loss**

Remember your mother warning you not to listen to loud music because you'd ruin your ears? She was right. In the U.S., one-third of people over sixty and half of those over seventy-five have hearing loss. The most common cause is long-term exposure to loud noises. Baby boomers are losing their hearing earlier than their parents and grandparents did, presumably because our worlds are noisier than they used to be. Some experts are particularly worried about portable MP3 players like the iPod, which can produce very loud music for hours without recharging.

It's not just rock and roll, of course. Hearing loss is caused by any loud noise that persists over time—a lawnmower, motorcycle, airplane, ambulance siren, or firecracker show. Even brief exposure to a very loud sound can damage your hearing. In these situations, where the noise isn't the point of the experience, you can protect yourself by wearing earplugs to keep the sound level down. A rock concert clocks in at the same noise intensity as a chainsaw—and experts recommend limiting exposure to those sounds to no more than one minute at a time. If you don't want to stop going to concerts, be aware that noise-induced damage is cumulative, so the more noise you experience over your life, the sooner you'll start to lose your hearing.

Noise causes hearing loss by damaging hair cells, which detect sounds in the inner ear. As discussed above, hair cells have a set of thin fibers called the hair bundle extending from their surface that move in response to sound vibrations. If the hair bundle moves too much, the fibers can tear, and that hair cell will no longer be able to detect sound. The hair cells that respond to high-pitched sounds (like a whistle) are most vulnerable and tend to be lost earlier than the hair cells that respond to low-pitched sounds (like a foghorn). That's why noise-related hearing loss tends to begin with difficulty in hearing high-pitched sounds. Sounds at this frequency are especially critical for understanding speech.

Ear infections are another common cause of hearing loss, so it's important to get them diagnosed and treated. Three out of four children get ear infections, and parents should watch for symptoms, which include tugging at the ears, balance or hearing problems, difficulty sleeping, and fluid draining from the ears.

least high-pitched sounds) coming from the right tend to be a little louder in your right ear; their intensity is reduced in your left ear because your head is in the way. (Low-pitched sounds can go over and around your head.) You use timing differences between your ears to localize low- and medium-pitched sounds and use loudness differences between your ears to localize high sounds.

Cocktail party: A gathering held to enable forty people to talk about themselves at the same time. The man who remains after the liquor is gone is the host.
—Fred Allen

When it's working to identify the content of a sound, the brain is specially tuned to detect signals that are important for behavior. Many higher brain areas respond best to complex sounds, which range from particular combinations of frequencies to the order of sounds in time to specific communication signals. Almost all animals have neurons that are specialized to detect sound signals that are important to them, like song for birds or echoes for bats. (Bats

Practical tip: Improving hearing with artificial ears

Hearing aids, which make sounds louder as they enter the ear, do not help patients whose deafness results from damage to the sound-sensing hair cells in the cochlea. However, many of these patients can benefit from a cochlear implant, which is an electronic device that is surgically implanted inside the ear. It picks up sounds using a microphone placed in the outer ear, then stimulates the auditory nerve, which sends sound information from the ear to the brain. About sixty thousand people around the world have a cochlear implant.

Compared to normal hearing, which uses fifteen thousand hair cells to sense sound information, cochlear implants are very crude devices, producing only a small number of different signals. This means that patients with these implants initially hear odd sounds that are nothing like those associated with normal hearing.

Fortunately, the brain is very smart about learning to interpret electrical stimulation correctly. It can take months to learn to understand what these signals mean, but about half of the patients eventually learn to discriminate speech without lipreading and can even talk on the phone. Many others find that their ability to read lips is improved by the extra information provided by their cochlear implants, although a few patients never learn to interpret the new signals and don't find the implants helpful at all. Children more than two years old can also receive implants and seem to do better at learning to use this new source of sound information than adults do, probably because the brain's ability to learn is strongest in childhood (see chapter 11).

Practical tip: **How to hear better on your cell phone in a loud room**

Talking on your cell phone in a noisy place is often a pain. If you're like us, you've probably tried to improve your ability to hear by putting your finger in your other ear but found that it doesn't work very well.

Don't give up. There is a way to hear better by using your brain's abilities. Counter-intuitively, the way to do it is to cover the mouthpiece. You will hear just as much noise around you, but you'll be able to hear your friend better. Try it. It works!

How can this be? The reason this trick works (and it will, on most normal phones, including cell phones) is that it takes advantage of your brain's ability to separate different signals from each other. It's a skill you often use in crowded and confusing situations; one name for it is the "cocktail party effect."

In a party, you often have to make out one voice and separate it from the others. But voices come from different directions and sound different from one another—high, low, nasal, baritone, the works. As it turns out, your brain shines in this situation. The simplest sketch of what your brain is doing looks like this:

voice ➤ left ear ➤ BRAIN ◄ right ear ◄ room noise

More complicated situations come up, such as multiple voices coming from different directions. The point is that brains are very good at what scientists call the source separation problem. This is a hard problem for most electronic circuitry. Distinguishing voices from each other is a feat that communications technology cannot replicate. But your brain does it effortlessly.

Enter your telephone. The phone makes the brain's task harder by feeding sounds from the room you're in through its circuitry and mixing them with the signal it gets from the other phone. So you get a situation that looks like this:

voice plus distorted room noise ➤ left ear ➤ BRAIN ◄ right ear ◄ room noise

This is a harder problem for your brain to solve because now your friend's transmitted voice and the room noise are both tinny and mixed together in one source. That's hard to unmix. By covering the mouthpiece, you can stop the mixing from happening and re-create the live cocktail party situation.

Of course, that brings up a new question: why do telephones do this in the first place? The reason is that decades ago, engineers found that mixing the caller's own voice with the received signal gives more of a feeling of talking live. The mixing of both voices—which is called "full duplex" by phone geeks—does do that, but in cases where the caller is in a noisy room, it makes the signal harder to hear. Until phone signals are as clear as live conversation, we are stuck with this problem—which you can now fix using the power of your brain. As the phone ad says, "Can you hear me now?"

use a type of sonar to navigate by bouncing sounds off of objects and judging how quickly they come back.) In humans, an especially important feature of sound interpretation is the recognition of speech, and several areas of the brain are devoted to this process.

Your brain changes its ability to recognize certain sounds based on your experiences with hearing. For instance, young children can recognize the sounds of all the languages of the world, but at around eighteen months of age, they start to lose the ability to distinguish sounds that are not used in their own language. This is why the English *r* and *l* sound the same to Japanese speakers, for instance. In Japanese, there is no distinction between these sounds.

You might guess that people just forget distinctions between sounds they haven't practiced, but that's not it. Electrical recordings from the brains of babies (made by putting electrodes on their skin) show that their brains are actually changing as they learn about the sounds of their native language. As babies become toddlers, their brains respond more to the sounds of their native language and less to other sounds.

Once this process is complete, the brain automatically places all the speech sounds that it hears into its familiar categories. For instance, your brain has a model of the perfect sound of the vowel *o*—and all the sounds that are close enough to that sound are heard as being the same, even though they may be composed of different frequencies and intensities.

As long as you're not trying to learn a new language, this specialization for your native language is useful, since it allows you to understand a variety of speakers in many noise conditions. The same word produced by two different speakers can contain very different frequencies and intensities, but your brain hears the sounds as being more alike than they really are, which makes the words easier to recognize. Speech recognition software, on the other hand, requires a quiet environment and has difficulty understanding speech produced by more than one person because it relies on the simple physical properties of speech sounds. This is another way that the brain does its job better than a computer. Personally, we're not going to be impressed with computers until they start creating their own languages and cultures.

CHAPTER 8

ACCOUNTING FOR TASTE (AND SMELL)

Animals are among the most sophisticated chemical detection machines in the world. We are able to distinguish thousands of smells, including (to name a few) baking bread, freshly washed hair, orange peels, cedar closets, chicken soup, and a New Jersey Turnpike rest stop in summer.

We are able to detect all these smells because our noses contain a vast array of molecules that bind to the chemicals that make up smells. Each of these molecules, called receptors, has its own preferences for which chemicals it can interact with. The receptors are made of proteins and sit in your olfactory epithelium, a membrane on the inside surface of your nose. There are hundreds of types of olfactory receptors, and any smell may activate up to dozens of them at once. When activated, these receptors send smell information along nerve fibers in the form of electrical impulses. Each nerve fiber has exactly one type of receptor, and as a result smell information is carried by thousands of "labeled lines" that go into your brain. A particular smell triggers activity in a combination of fibers. Your brain makes sense of these labeled lines by examining these patterns of activity.

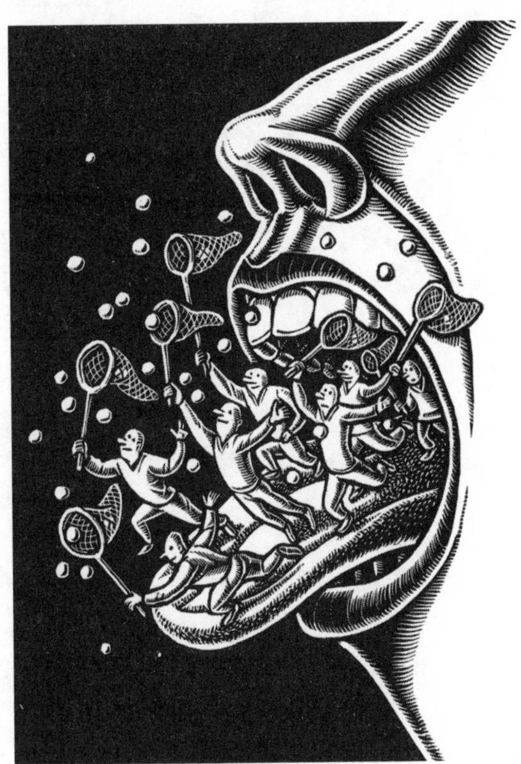

Did you know? A seizure of the nose, or sneezing at the sun

As many as one in four people in the U.S. sneeze when they look into bright light. This photic sneeze reflex appears to serve no biological purpose whatsoever. Why would we have such a reflex, and how does it work?

The basic function of a sneeze is fairly obvious. It expels substances or objects that are irritating your airways. Unlike coughs, sneezes are stereotyped actions, meaning that each occurrence of an individual person's sneeze follows the same course over time, without variation. The explosive beginning of a sneeze expels air at remarkable speeds, up to one hundred miles per hour. A powerful synchronized reproducible event like this can only be generated by positive feedback within some circuit somewhere in the brain, one that leads to a runaway burst of activity, reminiscent of the onset of epileptic seizures. However, sneezes are different in that they have a preset mechanism for ending, and they don't spread in an uncontrolled way to other bodily movements or activities.

The sneezing center is located in the brainstem, in a region called the lateral medulla; damage to this site causes us and other mammals to lose the ability to sneeze. Usually sneezing is triggered by news of an irritant that is sent through brain pathways and into the lateral medulla. This information comes from the nose to the brain through several nerves, including the trigeminal nerve, which carries a wide variety of signals from the face into the brainstem. The trigeminal nerves (we have one on each side) are cranial nerves with multiple functions: they process noxious and tactile stimuli from the face and much of the scalp, as well as from the conjunctiva and cornea of the eye. The trigeminal nerve even carries motor signals in the opposite direction, out of the brain, including the commands to bite, chew, and swallow; it's a really crowded nerve.

This crowded arrangement might explain why bright light could erroneously induce a sneeze. A bright light, which would normally be expected to trigger pupil contraction, might also spill over to neighboring sites, such as nerve fibers or neurons that carry nose-tickling sensations. Bright light isn't the only unexpected sensation that is known to trigger sneezes; male orgasm can also trigger sneezes (in the male who is having the orgasm).

Fundamentally, a crossed-wire phenomenon like the photic sneeze reflex is made possible because the circuitry of the brainstem is a jumbled, crowded mess. The brainstem contains critical circuitry for a wide variety of reflexes and actions, including almost everything our bodies do. The basic layout of the brainstem was worked out early in the history of the vertebrates. Thirteen pairs of cranial nerves are found in nearly all vertebrates (though fish have three additional pairs that carry signals such as those from the lateral line receptors along their sides). The cranial nerves lead to a complex network of specific clusters of neurons, or nuclei, which are basically arranged the same way and serve similar functions among all vertebrates. Indeed, looking at nervous systems in nonhuman animals is an extremely good way to guess at how the structures in our brains work.

The reason that brainstem structures are so similar across species is that the whole system is intricately constructed. From an evolutionary standpoint, it would be disastrous to move anything around on a wholesale basis. As successors to the earliest, simplest vertebrates, today's vertebrates (including fish, birds, lizards, and mammals) are all doomed to use a wiring layout that can be modified in small ways but not fundamentally changed. It's reminiscent of the New York City subway system, which was simple at one point, but is now hopelessly convoluted after planners added layer upon layer of complexity. Some parts of the brainstem aren't used any more, and the original core is now so jury-rigged and patched together, it can't possibly be replaced for fear of stopping the whole system cold. Frankly, the brainstem is about as good an argument against intelligent design as one could ever hope to find in nature.

Taste works the same way, except that flavor receptors are in your tongue. Taste is simpler since there are only five basic flavors: salty, sweet, sour, bitter, and umami. (What's umami, you say? It's the savory taste that's found in cooked meat or mushrooms or in the food additive monosodium glutamate, MSG. There's no word for it in English, which is why we use the Japanese term.) Each of these basic tastes has at least one receptor, sometimes more. Bitterness, for instance, is sensed by dozens of receptors. As animals evolved, they needed to detect toxic chemicals in their environments. Because toxic compounds came in many forms, it was necessary to have receptors that could detect all of them. This is why we have a natural repulsion to bitter flavors. This distaste can be overridden by experience; look at all the lovers of tonic water and coffee.

Why do we call spicy foods hot? The chemical that gives chili and hot sauce their zest is capsaicin. Your body also uses capsaicin receptors to detect warm temperatures. This is why you sweat when you eat spicy food—the receptors have what you might call a "hotline" into your brain to trigger responses to cool you off. You have capsaicin receptors not only in your tongue, but all over your body. One way to discover this is by cooking with hot peppers and then putting in your contact lenses. Ouch!

Did you know? **Why mice don't like Diet Coke**

The ingredient that makes Diet Coke sweet is aspartame (NutraSweet). It works by binding to sweet receptors in your tongue. In humans, the sweet receptor binds not only to sugar, but also aspartame, saccharin, and sucralose (Splenda). In mice, sweet receptors bind to sugar and saccharin, but not aspartame. They don't prefer water with NutraSweet to plain water, suggesting that to a mouse, Diet Coke wouldn't taste sweet. (It's a similar story for ants, which are not attracted by diet soda.)

Scientists have used genetic technology to replace the mouse's sweet receptor with the human sweet receptor. These transgenic mice like aspartame—and presumably Diet Coke. This proves that they use the same brain pathways to taste sweet things as we do, just with different receptors.

If you have pets, there's an experiment you can do at home. See how they like different kinds of sweet beverages—juice, sugared soda, and diet soda. Put out one dish of each and see what your pet goes for. You might be surprised at the results!

Minty foods taste cool for a similar reason. A receptor has recently been identified that binds to menthol. Plants may make menthol for the same reason that they make capsaicin—to make themselves taste bad to animals.

Smells and tastes often have strong emotional associations: your grandmother's apple pie, burning leaves, your lover's shirt, fresh coffee in the morning. Smells can also have negative associations. On September 11, 2001, and in the days after, Manhattan was permeated by a bitter, acrid smell that nobody who was there can ever forget. Some smells may be negative for some and positive for others. (Think of Kilgore's favorite smell in *Apocalypse Now*: "I love the smell of napalm in the morning . . . the whole hill smelled like victory.") These associations may occur because olfactory information has a direct connection into your limbic system, brain structures that mediate emotional responses. These structures are able to learn, raising the possibility that they allow you to associate smells with pleasurable or dangerous events.

CHAPTER 9

TOUCHING ALL THE BASES: YOUR SKIN'S SENSES

Pickpockets may not spend a lot of time talking about how the brain works, but their profession does require some practical knowledge of the subject. A common technique involves two partners in crime. One thief bumps into the victim on one side, to distract him from the other thief's hand taking something from the other side. This approach works because it draws the victim's attention to the wrong side of his body, which distracts his brain from events on the side where the important action is.

Expectations do not only influence our responses; they actually influence what we feel. Your perception of the body's sensations comes from the interaction of two processes: signals coming from receptors in your body, and activity in brain pathways that control your response to these signals—including, in some cases, whether they get passed along to the brain at all. This interaction is apparent not only in pickpocketing, but also in phenomena as diverse as pain and ticklishness.

Of course, the physical stimuli on your body also affect what you feel. Your skin contains a multitude of different receptors—specialized nerve endings that sense things like touch, vibration, pressure, skin tension, pain, and temperature. The brain knows which kind of sensor is activated, and where it is on the body, because each sensor has a "private line" that uses spikes to carry only one kind of information to the brain. Some parts of your body are more sensitive than others. The highest density of touch receptors is found on the fingertips, with the face a close second. Your fingers contain many more receptors than your elbows, which is why you don't explore an object with your elbow when you're trying to figure out what it is.

Another set of receptors in your muscles and joints gives you information about the positioning of your body and the tension in your muscles. This system is what allows you to be aware of the position of your arm when your eyes are shut. When these sensors are damaged,

Did you know? Why can't you tickle yourself?

When doctors examine a ticklish patient, they place the patient's hand over theirs during the exam to prevent the tickling sensation. Why does this work? Because no matter how ticklish you may be, you can't tickle yourself. Go ahead. Try it. The reason is that with every move you make, part of your brain is busy predicting the sensory consequences of that movement. This system keeps your senses focused on what's happening in the world so important signals aren't drowned out in the endless buzz of sensations caused by your own actions.

For instance, as we write, we are unaware of the feel of the chair and the texture of our socks. Yet we'd immediately notice a tap on the shoulder. If the only information your brain received was pure touch sensation, you wouldn't be able to tell whether someone was punching your shoulder or whether you'd just bumped into a wall. Since you'd want to react very differently to those two situations, it's important for your brain to be able to tell them apart effortlessly.

How does your brain accomplish this goal? To study this, scientists in London developed, of all things, a tickling machine. When a person presses a button, a robot arm brushes a piece of foam across the person's own hand. If the robot arm brushes the hand as soon as she presses the button to activate it, the person feels the sensation but it doesn't tickle. However, the effect can be enhanced by introducing a delay between the button press and the touch. A delay of one-fifth of a second is enough to fool the brain into thinking the robot's touch has been delivered by someone else—and then it tickles.

Even better, if the robot's touch is delivered in a different direction than the one in which the person pulls the lever, then a delay as short as one-tenth of a second is enough to generate a tickling sensation. This experiment shows that, at least for tickling, your brain is best at predicting the sensory outcome of a movement on the time scale of a fraction of a second.

So what happens in the brain when you try to tickle yourself? The same scientists used functional brain imaging, a technique that allowed them to observe how different parts of the brain respond to various types of touch. They looked at brain regions that normally respond to a touch to the arm. These regions responded when the experimenters delivered the touch. However, if someone delivered the touch to his own body, the response was much smaller—but still there. When the delay was increased, leading the touch to feel tickly, the brain responses became large once again. It's as if your brain is able to turn down the volume on sensations that are caused by your own movements.

This means that some brain region must be able to generate a signal that distinguishes your own touch from someone else's. The experimenters found one: the cerebellum. This part, whose name means "little brain," is about one-eighth of your total brain size—a little smaller than your fist—and weighs about a quarter-pound. It's also scientists' best candidate for the part of the brain that predicts the sensory consequences of your own actions.

The cerebellum is in an ideal location for distinguishing expected from unexpected sensations. It receives sensory information of nearly every type, including touch, vision, hearing, and taste. In addition, it receives a copy of all the movement commands sent out by the motor centers of the brain. For this reason, researchers suggest that the cerebellum uses the movement commands to make a prediction of the expected consequences of each movement. If this prediction matches the actual sensory information, then the brain knows it's safe to ignore the sensation because it's not important. If reality does not match the prediction, then something surprising has happened—and you might need to pay attention.

people find all kinds of movement to be very difficult, and they have to watch themselves as they move to avoid making mistakes.

As in other sensory systems, areas of the brain that analyze touch information are organized into maps, in this case, maps of the body surface. The size of a given brain area depends on the number of receptors in each part of the body, rather than on the size of that body part, so that the part of the brain's map that receives information from the face is larger than the area that receives information from the entire chest and legs. Along the same lines, in a cat's brain, a large area is occupied by neurons that respond to the whiskers.

Responses to painful stimuli are carried by separate receptors and analyzed by brain areas distinct from those that carry information about regular touch. One family of pain receptors detects heat and cold, while another family of receptors detects painful touch.

Practical tip: **Does acupuncture work?**

Having needles stuck into your skin doesn't sound like much fun, but a lot of people swear by it. The therapeutic use of needles, called acupuncture, is routine in Asia and has become increasingly common in the West over the past three decades. Roughly 3 percent of the U.S. population and 21 percent of the French population have tried it. About 25 percent of medical doctors in the U.S. and U.K. endorse acupuncture for some conditions.

The scientific evidence for medical benefits from acupuncture is mixed and very controversial. Many of the studies are done and evaluated by people with a vested interest in proving or disproving its effectiveness—making it difficult to know who you should listen to. In our reading of the scientific literature, the best evidence suggests that acupuncture is more effective than no treatment at all for some conditions, notably chronic pain and nausea. For most people, acupuncture seems to be about as effective as conventional treatments for these conditions, but there is little or no evidence that it's effective for other conditions, such as headache or drug addiction.

Traditional practitioners believe that acupuncture improves the flow of *qi*—a Chinese word that, roughly, means *energy*—circulating in pathways of the body. To unblock the energy flow, needles are inserted along these pathways, though different authors disagree on the exact locations, number of pathways, and acupuncture points. Attempts to identify these pathways in terms of the body's electrical or other physical properties have not been successful.

However, acupuncture definitely has some effect on the brain. Functional imaging of brain activity shows that acupuncture has specific effects on particular brain areas. For instance, an acupuncture point in the foot traditionally related to vision has been reported to activate the brain's visual cortex, while stimulation at other sites nearby do not. However, a follow-up study reported a different result, creating considerable uncertainty about this conclusion. Brain areas that control pain are activated by acupuncture—but also by the expectation of pain relief or by sham acupuncture at incorrect sites.

This brings up a major problem with evaluating any medical treatment (and especially acupuncture): a lot of patients feel better just because someone is paying attention to their problem. This is the reason that more than half of the patients in many studies report improvements in their conditions after taking sugar pills. Scientists solve this problem by doing double-blind studies, in which neither the patients nor the health care providers know who's receiving the real treatment and who's getting the fake one.

Of course, it's tough to keep patients guessing about whether needles are being stuck in them or not. Some researchers have used sham acupuncture, inserting needles into incorrect locations. Sham acupuncture is often found to be as effective as real acupuncture, but it's easy to believe that sham acupuncture might have some therapeutic effect in its own right. A few studies have used a telescoping probe that retracts as it approaches the skin, feeling like a needle to people who haven't experienced real acupuncture. This solves half the problem, but the practitioners still know whether they are giving real or fake treatments, which may lead them to behave differently with patients in the two groups and therefore influence their responses. Telescoping-probe studies have given mixed results. Most of them show real and sham acupuncture to be equally effective, but a large minority find real acupuncture to be more effective.

At the end of the day, you probably don't care why you're feeling better, as long as you are, and there's no reason not to try acupuncture if you're interested. In the hands of a qualified practitioner, it's pretty safe, causing serious problems for fewer than one in two thousand patients. Even if many of the details of the process turn out to be folklore, as we expect that they will, acupuncture does seem to have practical value in treating certain conditions.

Prediction is hard, especially of the future.

—Unknown

If you've ever touched a hot stove, you know that many pain receptors can activate reflex pathways that allow you to make a very rapid response to sensations that indicate the possibility of immediate danger to your body. However, these reflexes—and all responses to pain—are very strongly influenced by the person's interpretation of the painful situation. Indeed, there is an entire set of brain areas that influences activity in the direct pain-sensing parts of the brain based on context and expectation. This effect can be as powerful as a near-complete lack of pain in a soldier with a serious injury on the battlefield. More commonly, we've all seen the opposite effect—the sudden intensification of pain in a small child when his mother approaches.

These responses are often called psychological, but that doesn't mean they're not real: people's expectations and beliefs create physical changes in the brain. If people are given a pill or an injection that contains no active drug but are told that it will relieve their pain, activity increases in the parts of the brain that are normally involved in modulating pain. When people are told

Practical tip: **Referred pain**

Have you ever had pain caused by indigestion that made it feel like your chest was hurting? This sort of confusion happens because all the nerves that sense pain in the internal organs send signals through the same pathways in the spinal cord that carry information from the body surface. This convergence leaves the brain uncertain about what's wrong. Pain felt in a place other than its true source is called referred pain.

For this reason, doctors learn that when patients complain about pain in their left arm, it may indicate a heart attack. Similarly, pain from a kidney stone may feel like a stomachache, gallbladder pain may be felt near the collarbone, and pain from appendicitis may hurt near your bellybutton. If you have persistent pain without an apparent cause in any of these areas (but especially the left arm), you should see a doctor as soon as possible.

that a cream will reduce the pain of an upcoming electric shock or heat stimulus, they not only show increased activity in pain-controlling regions, they also show reduced activity in parts of the brain that receive pain signals. In addition, pain relief from such placebo treatments can be blocked by naloxone, a drug that prevents morphine from acting on its receptors. From these results, we can conclude that when patients are told that their pain will be reduced, the brain responds by releasing natural chemicals that reduce pain, which are called endorphins. Even a saltwater injection, the most innocuous treatment possible, can lead to pain relief—and also the release of endorphins.

Endorphins act on the same receptors that respond to morphine and heroin. The existence of endorphins is the reason that your body has receptors that respond to these drugs. Endorphins may allow pain relief when the brain decides that it's more important for the body to be able to go on (perhaps to escape from continuing danger) than it is to protect the injury from further damage.

Scientists at Stanford have been trying to use brain imaging to train people to activate pain-controlling areas of their own brains. If it works, this technique could allow people with chronic pain to reduce their own discomfort without needing fake pills or creams or injections. The scientists use functional imaging to detect activity in the target region of the brain. Subjects can see on a computer display whether they are achieving the desired effect. Using this technique, people have been able to gain voluntary control over the activity in one area of their brains—though it remains to be seen whether this approach will lead to pain relief in patients.

PART THREE

HOW YOUR BRAIN CHANGES THROUGHOUT LIFE

CHAPTER 10

GROWING GREAT BRAINS: EARLY CHILDHOOD

When we were kids, our parents tried to keep us safe and stop us from running with scissors. As far as we can remember, that was enough to keep their hands full. Today, middle-class family life has become a far more complicated affair. Daily activities are a blur of flash cards and baby aerobics. Magazines say that you can increase your children's intelligence by playing Mozart for them when they're young—or even before they're born. Parents worry that if little Emma doesn't attend the right preschool, she'll never get into a decent college. Every few years, another expert piles on more anxiety by explaining how a child's experiences in early life determine intelligence and success later on.

Our own parents had very different philosophies of child rearing. Sam spent hours each day watching television and can still recite the plot of almost every episode of *Star Trek* and *The Brady Bunch*. Sandra was five years old before friends at school let her in on the secret that there were other channels on TV besides PBS. Since her parents never watched anything else, she spent her early years with *Sesame Street* and other carefully designed educational fare. Yet Sam seems to have made up for any possible brain damage, and, indeed, as a university professor, is now even responsible for the training of younger minds.

It's true that the early environment influences how a child's brain grows, but you rarely need to worry that your child isn't getting enough stimulation. There's no question that childhood deprivation can interfere with brain development. To start with an extreme example, children who spent their early years in Romanian orphanages often have lifelong problems. But these poor kids were left alone in a crib for years, visited only by a caretaker who came along every so often to change diapers. Unless you're locking your kid in a closet (in which case,

Myth: Listening to Mozart makes babies smarter

One of the most persistent brain myths is that playing classical music to babies increases their intelligence. There's no scientific evidence for this idea, but it's proven amazingly persistent, probably because it allows parents to address their anxiety about their children's intellectual development—and because sellers of classical music for children encourage the belief every chance they get.

This myth began with a 1993 report in the scientific journal *Nature* that listening to a Mozart sonata improved the performance of college students on a complex spatial reasoning task. The researchers summarized the effect as equivalent to an eight- to nine-point gain on the Stanford-Binet IQ scale. Journalists didn't find this result immediately fascinating; they reported it about as much as any other science story published in the same journal that year.

The idea really took off after the 1997 publication of *The Mozart Effect* by Don Campbell, who brought mysticism together with loosely interpreted scientific results to produce a best-seller that influenced public policy. The next year, Georgia governor Zell Miller played Beethoven's "Ode to Joy" to the legislature and requested $105,000 to send classical music CDs to all parents of newborns in the state. The legislators approved his request, failing to notice that it made no sense to argue that music would lead to lifelong intelligence gains in babies based on an effect that lasts less than fifteen minutes in adults. Florida legislators soon followed suit, requiring state-funded day care centers to play classical music every day.

By now, the idea that classical music makes babies smarter has been repeated countless times in newspapers, magazines, and books. The idea is familiar to people in dozens of countries. In the retelling, stories about the Mozart effect have progressively replaced college students with children or babies. Some journalists assume that the work on college students applies to babies, but others are simply unaware of the original research.

In 1999, another group of scientists repeated the original experiment on college students and found that they could not duplicate its results. It hardly matters, though, that the first report had been incorrect. What's important is that no one has tested the idea on babies. Ever.

While playing classical music for your kids isn't likely to improve their brain development, something else will—having them play music for you. Children who learn to play a musical instrument have better spatial reasoning skills than those who don't take music lessons, maybe because music and spatial reasoning are processed by similar brain systems. Filling your house with music may indeed improve your children's intelligence—as long as they aren't passive consumers, but active producers.

you should stop doing that right away), you don't have to worry about how this sort of serious deprivation affects brain development.

You're probably more interested in how the brain grows under normal circumstances. The early stages of brain development don't require experience at all—which is fortunate, since they mostly occur inside the mother, where there's not a lot of stimulation available. This is when the different areas of the brain form, when neurons are born and migrate to their final positions, and when axons grow out to their intended targets. If this part of the process goes wrong, because of drugs or toxins in the mother's body or genetic mutations in the fetus, severe birth defects often result. Prenatal brain development is sufficient to permit many basic behaviors, such as withdrawal from a rapidly approaching object.

After the baby is born, sensory experience starts to become important for some aspects of brain development. In any normal environment, though, the odds-on bet is that most of the necessary experience is easily available. For example, as we saw in the story of Mike May (chapter 6), the visual system can't develop correctly without normal vision, but that experience happens effortlessly for anyone who can see. We don't have to send our kids to vision enrichment classes to make sure that these parts of their brains develop correctly. Scientists call this type of dependence on the environment "experience-expectant development," and it's by far the most common way that our experiences influence how our brains grow. Along the same lines, readily available sensory experience is necessary for the correct development of sound localization and for mother-infant bonding.

Sensory experience works by influencing which neurons receive synapses from the incoming axons. You might think that the patterns of activity in the growing axons would determine where new synapses are formed, but the brain doesn't take this approach. Instead, it produces a huge number of relatively nonselective connections between neurons in the appropriate brain areas during early development and then removes the ones that aren't being used enough over the first two years of life (in people). If the brain were a rosebush, early life experience would be the pruning system, not the fertilizer.

Experience-expectant development is also important for the development of a child's intelligence, as the effects of serious environmental deprivation show. There's some evidence that the ability to learn or to reason may also be enhanced by exposure to intellectually stimulating activities (what we often call enrichment), but exactly how much is a tricky question. A key element may be the difference between learning an active skill, such as playing a musical instrument, as opposed to passive exposure, such as listening to music (see *Myth: Listening to Mozart makes babies smarter*).

Over the past several decades, the average intelligence quotient (IQ) in many countries has increased, as we'll discuss further in chapter 15, suggesting that something about modern life is producing kids who do better on these tests than their parents. This effect is strongest among children with lower-than-average IQs. We don't know the degree to which these IQ gains in less intelligent children are attributable to changes in their intellectual environment versus better prenatal care and early childhood nutrition, though we would bet that all of these factors are important.

The evidence that environmental enrichment helps the brain has mainly come from research on laboratory animals. For instance, mice that are housed with other mice and an assortment of toys that are changed frequently have larger brains, larger neurons, more glial cells, and more synapses than mice that are housed alone in standard cages. The enriched animals also learn to complete a variety of tasks more easily. These changes occur not only in young mice, but also in adult and old mice.

Unfortunately, there's ambiguity about how to apply this work to people; we don't know how enriched we are compared to lab animals. Lab animals live in a very simplified environment; they rarely have to navigate through complicated places to search for food or find someone to mate with, and they certainly don't have to write college application essays. In practice, then, this research is not so much about the positive effects of enrichment on the brain but about the negative effects of deprivation in the typical laboratory environment. All this information together suggests that society should get a high return on investing in the enrichment of the lives of kids who are relatively deprived. Further improvements for kids whose lives are already enriched may do little or no good.

As we'll see in the next chapter, some aspects of brain development require very specific types of experience. The brain is not a blank slate but instead is predisposed at certain times

Did you know? Early life stress and adult vulnerability

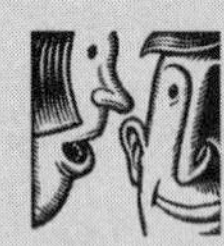

Some people just seem to be more mentally resilient than others. Part of the explanation may be that early experiences can increase the responsiveness of the stress hormone system in adulthood. This is true for rats, for monkeys, and probably for people as well.

In pregnant rodents, stress increases the release of glucocorticoid hormones. This hormone exposure can lead to a variety of later problems in the offspring. They're typically born smaller than normal animals and are more vulnerable to hypertension and high blood glucose as adults. These prenatally stressed animals grow up to exhibit more anxiety behaviors and are less able to learn in laboratory tests.

The good news is that getting a lot of maternal care in the first week of life can make rats less vulnerable to stress as adults. Maternal grooming permanently increases the expression of genes that encode stress hormone receptors in the hippocampus. Because activation of these receptors reduces the release of stress hormones, good mothering makes the rat pups less fearful later in life by reducing the responsiveness of their stress hormone system. Poor early mothering has the opposite effects. Artificially increasing the expression of these genes or housing the animals in an enriched environment reverses the hormonal effects of poor mothering in adult rats. Maternal grooming of pups also influences both excitatory and inhibitory neurotransmitter systems in adulthood.

Early stress may also increase vulnerability in humans. Abuse, neglect, or harsh, inconsistent discipline in early life increases the later risk of depression, anxiety, obesity, diabetes, hypertension, and heart disease. It also heightens the responsiveness of the stress hormone system in adulthood. However, scientists don't know whether stress causes changes in the brains of people similar to the changes observed in rats, nor do they know whether these effects might be reversible by drug treatments in adults.

to learn particular types of information. Although the specific language you speak as an adult depends on whether the people around you communicate in Swahili or Swedish, your brain is especially prepared to learn language at an early age. Your genes determine how you interact with your environment, including what you learn from it.

CHAPTER 11

Growing Up: Sensitive Periods and Language

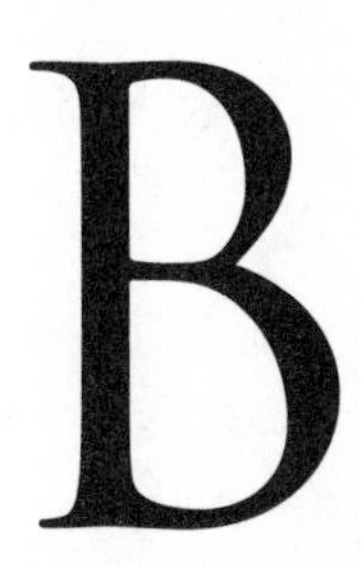

abies are incredible learning machines. You probably know that there's something unique about young brains when it comes to learning. But you may not appreciate that their abilities are very specific. Babies aren't sponges waiting to soak up anything that happens to them. They come into the world with brains that are prepared to seek out certain experiences at defined developmental stages.

The times early in life when experience (or deprivation) has a strong or permanent effect on the brain are called sensitive periods in development. They're the reason that people who learn a language as adults are more likely to speak with an accent. People can still learn when they're older, of course, but they learn some things less thoroughly, or in a different way. On the other hand, many types of learning are equally easy throughout life. There's no special advantage to being young if you want to study law or learn how to knit, but if you want to be a really good skier or speak a language like a native, it's best to learn as a child.

In some ways, sensitive periods can be thought of as being analogous to constructing a home. When you are building a house, you decide how you want to arrange the bedrooms. Once the house has been built, changes are much harder. You can rearrange or replace the furniture, but unless you're willing to put in an awful lot of work, your floor plan is set.

In the same way, the mechanisms by which brains are built allow some changes to be made much more easily early in life. Although learning is easier during sensitive periods, it can often still take place later on. Some people succeed in mastering a second language so that they sound indistinguishable from native speakers. Even if a language learned in adulthood is completely fluent, though, brain imaging shows that different, nearby parts of the brain are active when people hear their two languages. So not only are children better at learning languages than

Did you know? Is language innate?

We can't deny that learning is important for language development—after all, Chinese babies adopted by American parents grow up speaking English, not Mandarin—but an influential theory suggests that the brain is not infinitely flexible about what types of language it can learn. Instead, people seem to be constrained by a set of basic rules for constructing sentences that is hardwired into the brain.

The universal grammar idea was originally proposed by the linguist Noam Chomsky, who said that the languages of the world are not as different as they seem on the surface. The vocabulary may vary enormously from one language to the next, but there is a relatively limited range of possibilities for how sentences may be constructed. In this view, the grammar of a particular language is thought to be defined by a few dozen parameters, such as whether adjectives are placed before the noun, as in English, or after the noun, as in Spanish. It's as if babies learn to flip switches for various parameters in their brains, yielding the full grammatical complexity of a language from a small number of simple instructions.

Linguists have searched through the languages of the world, cataloging differences and similarities in an attempt to define these parameters. This is slow work, in part because many languages are related to one another. For example, French, Spanish, and Italian are Romance languages, which have similar-sounding vocabularies because they are descended from the same older language. For this reason, the best examples to test the universal grammar hypothesis are the most unusual languages, which are least related to the world's major languages and thus hardest for scientists to identify and analyze.

More reliable support for this idea comes from attempts to teach people artificial languages that don't follow the universal grammar rules. For example, various educators of deaf children have tried to invent new sign languages that are closer to the local spoken language. Most such languages do not follow the rules of universal grammar, and the children do not learn them well. What commonly happens is that children learn the language "incorrectly"—changing it to conform to universal grammar rather than accepting the artificial language as presented by the teacher.

adults, but they can also use a single area of their brains to support several languages. It's as if, to support new language acquisition, adults have to expand into the spare room.

When does the sensitive period for learning a native language end? This question has been hard to answer because almost all children are exposed to language early in life; if they aren't, they've usually been abused in other ways as well. However, one group—deaf children—often learns language late in the context of a normal life, and so has been much studied.

Deaf children are almost always born to hearing parents, and some of these children don't start learning sign language until they go to school. Some deaf children don't meet anyone who knows sign language until adolescence or beyond. When they do learn language, they use gestures rather than sounds. Despite the use of gestures, sign language is very comparable to spoken language. Sign language has grammar; for instance, American Sign Language has a grammar that is similar not to spoken English but to Navajo. Just like spoken language, sign language isn't one language, but a group of quite different languages. A deaf person from Britain would have no luck communicating with a deaf person from the U.S. unless one of them had learned the other's sign language, although the two countries share a spoken language.

Signed and spoken languages use similar brain mechanisms. They both involve the same language areas, which are in the left hemisphere in 97 percent of people. Broca's area in the frontal lobe of the cortex is responsible for producing language, and Wernicke's area in the temporal lobe is responsible for understanding it. Signed languages also have emotional tone, which in speech is called prosody. Prosody is generated in regions of the right hemisphere that correspond to Broca's and Wernicke's areas. The two types of languages follow similar grammatical rules (see *Did you know? Is language innate?*), and there's even a sign language equivalent of an accent in which poor speakers consistently get their finger and hand shapes a bit wrong. So signed and spoken languages have deep similarities, suggesting that studies of late learners of sign language can give us valid information on the limits of spoken-language learning.

As expected, children who learn sign language when they're younger are more fluent than children who learn when they're older. Up to age seven or eight, children are able to learn additional languages, spoken or signed, without any noticeable problems. Children who learn after the age of twelve almost never end up using sign language fluently; typically they have poor grammar and an accent (see above). Between these ages, there's a lot of individual variation in how well children learn sign language.

Some of the kids who learn at intermediate ages have accented but grammatically correct sign language. Similarly, hearing children retain the ability to pronounce sounds like a native speaker until some time in elementary school. The ability to learn grammatical rules seems to extend even further, maybe into junior high school. At some point, though, almost everyone reaches an age after which any new language will be learned as a second language.

The range of ages at which different language skills become less plastic illustrates another important point about sensitive periods: their timing is different for different types of learning. The time window for learning the sounds of a language happens earlier than the window for

Did you know? Is music like a language?

Both music and language involve elements arranged into sequences that are variable but follow certain rules. This similarity led scientists to consider whether the brain might process these two types of information in the same way. So far, the verdict is mixed. Functional imaging shows that tasks involving musical harmony activate Broca's area, which is necessary for speech, and a corresponding area in the right hemisphere that is important for prosody (intonation, which tells a listener when you're being sarcastic, for example, or asking a question). Music and language also both activate brain areas involved in the timing of auditory information. However, people with brain damage can lose their language abilities without losing their musical abilities, and vice versa, so these two functions are at least partly separated in the brain. There's no reason that this question needs to have a yes or no answer: it's likely that the brain areas that process language overlap partially, but not completely, with those that process music.

If this view is correct, it might offer a scientific basis for the widely held belief that childhood training is necessary for the achievement of high musical skill. Some aspects of auditory development do benefit from experience. In laboratory animals, the map of sound frequency in the auditory cortex requires normal experience during a sensitive period. In people, responses to tones do not become adultlike until the age of twelve or so. In deaf people, these responses remain abnormal into adulthood. Pitch perception also is more easily learned in childhood. Absolute pitch (the ability to recognize tones in isolation, instead of by their relationship to other notes) seems to require both a genetic predisposition and relevant auditory experience before age six. Absolute pitch is more common among people who speak tonal languages, such as Chinese, in which pitch is important for distinguishing words.

Is there a sensitive period for musical training in particular? The brains of adult professional musicians and nonmusicians differ anatomically, but this could be due to genetic differences. Musicians' brains also have different electrical responses, which are specific to the notes produced by their own instruments, and so probably result from experience. Some of these effects are more pronounced in musicians whose training began earlier in childhood, before age ten, and harmonic structure is thought to be easier to learn before age eight. All in all, we'd bet that musical training in early life does pack some extra punch, but later training has at least some effect as well. Stravinsky, for example, trained as a lawyer and didn't begin composing until he was twenty.

learning grammar. Along the same lines, the ability to see motion appears to develop earlier than the ability to see objects (see chapter 6). That means there's no such thing as a single broad sensitive period—only specific sensitive periods for particular types of learning.

Fortunately, society places limits on the experiments that can be done with babies, so scientists have turned to other species for insights into the biology of sensitive periods. Songbirds like zebra finches, for instance, have to learn their individual tunes from other birds, usually their fathers. If a young male has no one to learn from, he'll end up with a weird-sounding song that won't help him attract a mate as an adult.

Like babies, young songbirds are not infinitely flexible in what they can learn. Zebra finches raised by a closely related species, the Bengalese finch, do not learn the Bengalese song correctly. In some cases, a zebra finch will copy a few sounds from his foster father's song, but he will put these sounds into a typical zebra finch note sequence, which seems to be innate.

You may think you suffer from information overload in your daily life, but imagine what it's like to be a newborn. Without some way to separate the relevant from the irrelevant stimuli, babies might spend their energy learning to imitate the sounds of birds, or the washer and dryer, which would lead them to a very strange social life when they grew up. Luckily for all of us, the brain does not come into the world as a blank slate after all, but has its own firm ideas about what it should be learning.

CHAPTER 12

REBELS AND THEIR CAUSES: CHILDHOOD AND ADOLESCENCE

We like to think of ourselves as sober, responsible adults—gainfully employed, settled down, that kind of thing. But we were not always such upstanding citizens. Between the ages of thirteen and twenty-three, we had no fewer than five car accidents and three trips to emergency rooms between us. All these events were probably at least somewhat preventable, given that our lives have been far less dramatic since that period. Fortunately, we arrived in adulthood more or less intact—and able to write about what our brains were up to during that stormy time.

During adolescence, brains and bodies undergo great changes that accompany the transition into adulthood. This transition can include attaining greater independence from parents, taking on responsibilities such as a job and a family, and going through periods of emotional turmoil. This last type of transition is likely to be driven by changes in the brain. Young adults across a variety of mammalian species, including humans, have poor impulse control and are more likely to take risks, compared to younger animals or adults. Around puberty, many mammals also become more focused on social interaction and place a high value on novelty.

These changes may be a consequence of the late formation of some brain systems in youth. Over the course of adolescence, young adults show improvements in the planning and organization of behavior, response inhibition, attentional capacity, memory, and emotional self-control, suggesting that these systems are still developing. Although the brain has reached 90 percent of its adult size by the age of six, during the last 10 percent of growth, a lot is going on. Connections are being formed rapidly, but different brain regions develop at different rates. Some of

Practical tip: Improving your brain with video games

Instant messaging, cell phones, e-mail, TV, video games, animated billboards—the modern world is full of nonstop action, and it all seems to be happening at once. If you're over thirty, you've probably wondered why younger people aren't overwhelmed by all this stimulation.

The reason is that their brains are trained to handle it. Sustained practice at multitasking increases one's ability to pay attention to many things at the same time. A major source of practice is playing action video games—you know, the kind that parents hate, where the aim is to shoot as many enemies as possible before they shoot you. These games require players to distribute attention across the screen and quickly detect and react to events. Unfortunately, playing Tetris doesn't have the same effect on the brain, perhaps because it requires players to concentrate on only one object at a time, rather than multitasking.

In one study, college students who played action games regularly could count 50 percent more items in a very brief visual stimulus than students who didn't play. The players also processed information more quickly, could track more objects at once, and had better task-switching abilities. You might imagine that people with naturally strong abilities were better at the games and thus chose to play more often. But a group of nonplayers was able to improve their attentional capacity after training one hour per day for ten days on an action game, suggesting that these skills develop as a direct result of practice.

Does this mean that parents should encourage their kids to play shoot-'em-up action games? We wouldn't go out of our way to expose kids to violent images, but at least parents can take heart that video game playing has positive effects. In the long run, we'd love to see somebody make a lot of money by designing action-based video games that motivate kids to practice multitasking and improve their attentional capacity without using violence as the motivator. Kind of like Sim City . . . on a runaway bus!

the last connections to be formed are in the prefrontal cortex, a brain region that is important in moral reasoning and planning for the future. Adolescents may be only partway down the road toward having a fully functional set of prefrontal connections.

Another possible explanation for adolescent behavior derives from work on rodents, so we don't yet know if it applies to humans. Neurons containing the neurotransmitter dopamine, as well as the sensitivity of their targets, may help set individual levels of risk taking and responsiveness to rewards, which include social experiences, novelty, and psychoactive drugs. These neurons connect to the prefrontal cortex as well as the striatum and areas important for processing emotions, such as the nucleus accumbens and the amygdala. The balance between these connections seems to change during adolescence. In the early stages, cortical connections dominate and others are weaker, which seems to favor novelty seeking; this situation reverses by late adolescence. During adolescence, the cortical dopamine system is thought to be particularly sensitive to stress, making animals—both human and rodent—more vulnerable to stressors.

The brain maturation process also seems to make teenagers vulnerable for the first time to a variety of psychiatric disorders. Adolescence is marked by a gradual increase in the risk of mood disorders and psychosis, as well the emergence of gender differences in these disorders. People who are diagnosed with schizophrenia in their twenties often turn out to have exhibited their initial symptoms in adolescence. Similarly, rates of depression and anxiety disorders begin to increase at thirteen or fourteen years of age and reach adult levels by age eighteen. Twice as many women as men suffer from these mood disorders, and this difference emerges at puberty. Why puberty increases the risk of such brain dysfunction is not well understood at this point.

We still have only a preliminary understanding of how brain structures generate behaviors. Although prefrontal brain structures are still developing at times when risk taking and impulsivity are high, it isn't clear when or how a partially mature brain structure begins to function. For instance, prefrontal development does not appear to be very different in men and women, yet men engage in far more risky behaviors. (Our experience follows suit: of all our scrapes, only one car accident and one emergency room trip were Sandra's; the rest were Sam's.) The basis of this gender difference is not clear, though it may be related to differences in dopaminergic systems, as male rats show a much stronger decrease in dopamine receptors in the striatum than females during adolescence.

The idea that delayed brain maturation explains adolescent behavior is an attractive one and has been discussed extensively by journalists. With adolescence commonly characterized by rebellion, risk taking, and a tendency to ignore consequences, it's no surprise that parents are excited by research suggesting that teenage brains are not yet fully formed. It's comforting to think that bad behavior is a result of delayed brain maturation. That means it's not the parents'

Did you know? Brain growth and intelligence

You might imagine that a bigger brain would be associated with greater intelligence, but the relationship between brain size and intelligence is weak in adults, and there's no measurable relationship in young children. However, research suggests that intelligence and brain structure could be related in a more subtle way during brain development.

A key component of intelligence may depend on when in development synapses are formed—and removed. One study found evidence that intelligence correlates with patterns of growth and shrinkage during childhood and adolescence. Over the course of more than a decade, the scientists used imaging methods to monitor the brain structures of more than three hundred children, tracking their development from age seven to nineteen. They divided the kids into three groups according to how they did on a standardized IQ test.

Higher intelligence was related to the timing of the cortical sheet's thickening: the higher a child's intelligence, the later the thickness of his or her cortex peaked. The thickness of the cortical sheet ended up the same in all three groups by the age of nineteen. On average, the sheet thickness peaked earliest in children of average intelligence, and latest in children with IQ scores greater than 120. Peak thickness, after which shrinkage to adult levels occurred, typically began between age seven and nine in normal or above-average children, but was delayed until age eleven in the highest-IQ children.

What is happening in the brain during these changes? It's not the birth of new neurons. The brain reaches approximately 90 percent of its adult size by age six, when nearly all the neurons of the brain have already been born. The remaining increase in brain size has to be caused by other forms of growth. For instance, dendrites and axons run through the thickness of the cortex, suggesting that they might become longer or more bushy at a prolonged, steady pace in high-achieving kids. Increases and decreases in cortical thickness may therefore be related to the formation and loss of synaptic connections.

The growth and shrinkage of synaptic connections is interesting because it suggests that the formation and weeding out of connections between neurons might be critical aspects of intellectual development in children and teenagers. But even though these differences are starting to be noticed between groups of children, it's not time to send your child for a brain scan. All the trends we've described were only apparent by averaging the results from dozens of children. The effects are too small to predict how your child will do in school.

fault, it's not the kids' fault, and most importantly, it's a problem that will solve itself as they grow up.

Although the evidence that delayed brain maturation is responsible for adolescent behavior is largely speculative, the idea does have some support. One aspect of brain structure continues to develop until about age twenty-one: long-distance connections. Although most neurons are present by age two, the connections between them take much longer to mature. Axons, the wires that carry electrical signals from one neuron to another, are covered with an insulating sheath called myelin that allows electrical signals to move faster and more efficiently. The process of myelination is the last stage of brain development, and it's not complete until early adulthood. The last brain area to finish myelinating is the prefrontal cortex, which is important for inhibiting behavior and selecting behavior that's appropriate for meeting goals—two abilities that many teenagers seem to lack. At the same time, emotional areas are fully developed. This discontinuity in development may mean that emotions aren't regulated as well as they should be.

Even though prefrontal areas are still growing at this stage, other brain regions have developed to adult levels of size and myelination. As a result, adolescents are mature in their reflexes and their capacity to acquire new information. Indeed, compared to adults, they learn—and forget—new facts more quickly.

All these signs of maturity and aptitude can make young people highly functional. Indeed, many rural cultures across the world begin to treat young people as adults when they are twelve or thirteen years old. To a modern reader it may seem odd, but adolescence is a relatively recent invention, largely restricted to urban societies within the past century or so. This could be due to the increased complexity of life in the twentieth and twenty-first centuries, which requires that education last longer. Or perhaps growing up, like so many tasks, has expanded to fill our now longer lifespans.

CHAPTER 13

An Educational Tour: Learning

Imagine a dog who hangs around the front yard and chases every car that comes down the street. One day, a red Corvette driven by a neighborhood teenager hits the dog, fracturing its leg. The dog's owner would like this experience to teach his pet the lesson that chasing cars is a bad idea. But that's not the only possibility. Instead the dog may learn that he shouldn't chase red cars, or that he should go to another street to chase cars, or that he should be afraid of teenagers. On the other hand, imagine another dog who has been beaten by his first owner and now remains forever afraid of people, no matter how kind they are. The first dog has not generalized enough from his experience, while the second dog has generalized too much.

We all learn from experience, but figuring out exactly what we should learn can be very complicated. We all know people who repeat the same mistake over and over, even though they're punished for it, or who decide from one bad relationship that no potential partner can ever be trusted. Why does this happen?

What we learn is influenced by many factors: the biological characteristics of our species, individual genetic factors, and personal experience. Not only do different animals have different natural behaviors, they're also specialized to learn certain behaviors more easily than others. Animal trainers know that it's easy to teach tricks that follow these natural tendencies, but very difficult to go against them. In the wild, pigs make their living by digging up roots with their wide, flat noses. Not only are their bodies shaped by evolution to suit this activity, but so too are their brains. For this reason, it's hard to teach pigs to balance a coin on their noses; instead, the pigs tend to bury the coin and dig it up repeatedly, even if this activity is unrewarded or punished. Similarly, chickens tend to peck at things, so it's easy to train them to peck a key for a

Practical tip: Should you cram for an exam?

We've all done our share of cramming. Nearly everyone gets into a situation at some point where they've fallen behind in class and there's not enough time to catch up before the test. Studying intensively at the last minute may allow you to pass the exam, which certainly has some value, but it's not the best use of your time. Why? Psychologists have known for more than a century that your brain retains many kinds of information longer if it has an opportunity to process what you've learned between training sessions.

The advantage of spread-out learning is large and reliable. Two study sessions with time between them can result in twice as much learning as a single study session of the same total length. Spaced training works with students of all ages and ability levels, across a variety of topics and teaching procedures. Unsurprisingly, it also works with other animals, so you'd do well to remember this principle when you're trying to train your dog.

reward, but hard to teach them to stand on a platform without scratching or pecking. Some behaviors can't be conditioned at all. For instance, rewarding a hamster for scratching herself is an exercise in futility; hamsters will scratch only when they feel like it, no matter how hard you try to persuade them to change their ways.

Learning also varies among individuals of the same species. Behavioral differences between individuals are mainly due to differences in their brain anatomy, particularly in the connections between neurons. Are you an impulsive person who reacts quickly to events, or is your behavior calm and deliberate? Are you a talented skier? Do you know the capitals of all fifty states? Are you good at solving mechanical problems? All these abilities are based on the way that your neurons talk to each other, a combination of how your brain was wired up when you were a baby and the connections that have formed or broken since then through learning.

Neural connections generally follow a rule known to your high school coach: use it or lose it. Neurons strengthen synapses that are effective and weaken or remove synapses that stay silent when other synapses are being used. This process occurs more easily in babies, but it continues throughout adult life. Every day, your kids come home from school—or from basketball practice—with brains whose neurons are connected slightly differently from when they woke up in the morning.

Remember from chapter 3 that when an electrical signal arrives at the end of an axon, it triggers the release of a chemical neurotransmitter, which binds to receptors on the neuron at

Did you know? **Why are some things easier to learn than others?**

Sooner or later, most people discover that a single experience can lead to intense and sometimes permanent learned responses. For Sandra, it's orange juice, which didn't taste good again for years after the unfortunate college party where it was mixed with excessive amounts of vodka. For you, it might be the shellfish that you can't eat anymore, ever since you ran across that bad oyster at lunch last year. Taste aversion is a vivid example of prepared learning. It's easy to develop an intense dislike of what you ate before you got sick, even if it happened only once, but you never hear anyone say, "I can't bear to look at the shirt that my date was wearing the night I was ill." Logically, this makes sense because fashion choices are unlikely to make you physically sick (though fashionistas might make a few exceptions to this rule).

Many types of illness are caused by food. How does the brain know that food has a special connection to illness? We said in chapter 10 that babies' brains are not sponges waiting to soak up anything that happens to them. It probably comes as no surprise to hear that adults also have distinct predispositions for learning. Many of these tendencies—to learn some things easily and others not at all—seem to be built in at birth, in humans and other animals. Because evolution selects for outcomes, this approach can be an efficient way to make sure that an animal is a good fit for its environment, especially when the details of the surroundings can't be predicted in advance.

the other side of the synapse. In most cases, multiple synapses need to be active at once to trigger an action potential in the next neuron in line. When this happens, all the active synapses are strengthened so that they will have more influence on the recipient neuron the next time, either by releasing more neurotransmitter or by having more receptors available to receive the signal. This strengthening process is called long-term potentiation or LTP. At most synapses, the rule for inducing LTP is similar to one of the common behavioral rules for learning: stimuli will be associated if they occur close to the same time. In neurons, by analogy, synapses will be strengthened if they are active at the same time, which is often the result of receiving two stimuli simultaneously.

Of course, synapses can't be strengthened indefinitely, or eventually they'd all be maxed out, and the brain would lose its ability to learn new information. There are a few tricks that the brain uses to avoid this problem, but the most straightforward is a use-dependent weakening of synaptic connections called long-term depression or LTD. Synapses are weakened if they become active at a time when the recipient neuron isn't receiving enough stimulation to fire an action potential. Another trick is that on long timescales, new synapses can form and old ones can go away, which allows connections to be redistributed.

These changes, which are collectively called synaptic plasticity, occur more easily at certain times, such as infancy. In adults, synaptic plasticity comes more easily to particular parts of the brain such as the hippocampus, which we'll discuss further in chapter 23. Your brain has about a dozen different known ways of learning information, each of which uses a somewhat different combination of brain regions. For example, learning new facts and places causes changes in your hippocampus and cortex, while learning a new dance step changes your cerebellum.

Researchers know a lot about the signaling pathways and molecules that are involved in synaptic plasticity. Scientists have been able to use this knowledge to produce mice that find it harder or easier to learn, simply because they are missing a single gene from their DNA. This work suggests that modifying synapses is one of the brain's most important jobs. There are literally hundreds of genes that affect learning and dozens that affect overall intelligence. Many pathways perform similar jobs and can substitute for one another if the need arises, giving a measure of protection against the complete failure of learning, which would be devastating to an animal.

A particularly well-understood and important type of learning is fear conditioning, the process of learning to become afraid of stimuli in the environment that predict bad things are about to happen. A common type of fear-conditioning experiment goes like this: a rat is placed in an unfamiliar cage, a tone comes on, and then the animal receives a

Practical tip: Put it out of your mind

Practice makes perfect, or so we hear. Many elite performers, from athletes to actors, learn to start their training by mentally rehearsing the outcomes that they would like to achieve. Repeated visualization of a desired experience can be a very effective way to create a strong mental image in your brain.

Unfortunately, a lot of people end up using essentially this same rehearsal strategy when remembering bad experiences. It's unintentional, of course, but the effect of mentally rehearsing an experience over and over is the same, whether you're deliberately trying to increase the intensity of the memory or just doing it by accident because you're naturally inclined to think about the bad things that happen to you. Some doctors think that post-traumatic stress disorder, which we discuss in chapter 17, is partly caused by this sort of mental rehearsal.

The best strategy is easy to state. To develop a strong mental image of something you want to accomplish, visualize it repeatedly in as much detail as you can. If something is making you unhappy and you want to get it out of your head, try not to think about it too much. This is especially true of things that make you afraid.

This strategy can be hard to achieve in practice. One approach to try is to distract yourself. The approach can be direct: some psychologists recommend wearing a rubber band around your wrist and snapping it every time the persistent thought enters your head. Or it can simply involve doing something that you find engaging, whether that's playing a team sport or listening to music or going to the races. It will probably help to tell your friends and family that you've decided not to dwell on the problem anymore and ask them to remind you of that decision if you bring up the subject again. Then go do something productive or fun, as long as it's challenging. If the intrusive thought persists, it might be time to see a therapist, as we discuss in chapter 17.

mild electric shock. After a few experiences of this sort, the rat learns to anticipate the shock by freezing (a typical rodent fear response) whenever it hears the tone.

Scientists at New York University showed that auditory signals travel directly from the thalamus to the amygdala, a small region that is important for emotional responses, particularly fear. Neurons in a particular region of the amygdala fire more action potentials in response to the tone after conditioning than they did before the animal had learned to fear the tone. These changes in the electrical responses of neurons occur around the time that the animals start to show fear behavior, suggesting that they may cause fear-induced learned freezing. Similarly, rats or people with damage to the amygdala don't form normal fear memories.

Fear conditioning can be counteracted by a process called extinction, which is induced by repeatedly exposing a conditioned animal to the tone without the electrical shock. If this happens often enough, the animal will learn to stop freezing when it hears the tone, and the amygdala neurons will stop firing so strongly in response to the tone as well. However, extinction is a second form of learning that is overlaid onto the original fear conditioning; it does not restore the brain to its original state. Extinction seems to involve learning in the prefrontal cortex, a brain region that selects appropriate behaviors in context. Neurons in the prefrontal cortex become more active after extinction training, and they then suppress the activity of the amygdala neurons in response to the tone. Rats with damage to the prefrontal cortex can learn to fear a tone, but although extinction reduces their responses temporarily, the learning doesn't last, so the next day they act as though the extinction training never happened. Like other types of learning, extinction is influenced by an animal's natural tendencies. It is much more difficult to extinguish fear of stimuli, like snakes or spiders, that were important sources of danger in the evolutionary history of our species.

The amygdala also mediates the effects of emotions on other types of learning. Emotional arousal facilitates attention to the most important details of an experience. For instance, victims of armed robbery usually remember what the gun looked like. Patients with amygdala damage, on the other hand, may fail to concentrate on relevant details at stressful moments. In rats and humans, transient stress enhances learning in two ways: via the release of adrenaline and glucocorticoids. Both hormones act on receptors in the amygdala and hippocampus to enhance synaptic plasticity. However, chronic stress can greatly impair learning ability. This is another fact that's worth remembering when you're trying to train your dog.

Each of the brain's different learning systems has its own special properties. In the case of fear conditioning, the amygdala system allows you to learn one-time occurrences if they induce enough fear. At the other extreme, consider the number of repetitions necessary for most people to remember long lists of facts, an incredibly dull task that uses a different system, the hippocampus.

Most tricks for learning facts take advantage of the natural ways that human beings learn. Just as pigs gravitate toward burying things and chickens tend to peck, we have ways of learning in the natural world that come more easily to us. As described in chapter 6, humans are exceptionally visual animals, and at least one-third of our cortex works with visual information in one form or another. In addition, sequences of events and the physical proximity of objects to one another are natural groupings for us, since these are the ways that we experience the world. The hippocampus handles both the learning of facts as well as the learning of events and sequences.

One effective strategy combines several of these tricks: imagine that you are walking through a house, and each fact that you want to remember is associated with a particular place in the house. If this seems like a tedious business, you could achieve single-trial learning with the amygdala system. Unfortunately, this would require you to experience intense fear with every fact that you learn. It's not worth it.

CHAPTER 14

Reaching the Top of the Mountain: Aging

We hadn't been paying much attention to the research on aging and how to improve our chances of keeping our brains healthy for as long as possible. Now we're glad we wrote this book, because it's time for us to make some lifestyle changes that should make our retirement years happier.

Let's start with the bad news. Even putting aside diseases of aging like dementia, your brain's performance is likely to get worse as you get older. There are two main problem areas. The one that everybody knows about is memory. You may have more trouble keeping track of your car keys than you used to; this ability starts to deteriorate in your thirties, on average, and continues to decline as you age. Spatial navigation relies on a part of the brain involved in memory, the hippocampus, and this ability is also impaired with age in many animals, including humans.

The other problem area is what scientists call "executive function," which is the set of abilities that allows you to select behavior that's appropriate to the situation, inhibit inappropriate behavior, and focus on the task at hand in spite of distractions. Problems with executive function begin later, for most people when they reach their seventies, and include the deterioration of basic functions like processing speed, response speed, and working memory, the type that allows us to remember phone numbers for long enough to dial them. Difficulties with executive function, along with navigation problems, explain why your grandfather doesn't drive as well as he used to. (It's probably just as well that he can't remember where he put his car keys.) Some sensory inputs decline with age, like the hearing problems we discussed in chapter 7. It also gets harder to control your muscles, though it's not clear whether this problem lies with the brain or with the aging body.

Practical tip: How can you protect your brain as you get older?

The most effective approach to keeping your brain healthy with age turns out to be something you probably wouldn't expect: physical exercise. Neurons need a lot of support to do their jobs correctly, and problems with an aging circulatory system can reduce the blood supply that brings oxygen and glucose to your brain. Regular exercise, of the type that elevates your heart rate, is the single most useful thing you can do to maintain your cognitive abilities later in life.

Elderly people who have been athletic all their lives are much better at executive-function tasks than sedentary people of the same age. This relationship could occur because people who are healthier tend to be more active, but that's not it. When inactive people get more exercise, even in their seventies, their executive function improves in just a few months. To be effective, exercise needs to last more than thirty minutes per session and occur several times a week, but it doesn't need to be extremely strenuous. (Fast walking works fine.) The benefits of exercise seem to be strongest for women, though men also show significant gains.

How does exercise help the brain? There are several possibilities, all of which could contribute to the effect. In people, fitness training slows the decline in cortical volume with age. In laboratory animals, exercise increases the number of small blood vessels (capillaries) in the brain, which would improve the availability of oxygen and glucose to neurons. Exercise also causes the release of growth factors, proteins that support the growth of dendrites and synapses, increase synaptic plasticity, and increase the birth of new neurons in the hippocampus. Any of these effects might improve cognitive performance, though it's not known which ones are most important.

Beyond normal aging, exercise is also strongly associated with reduced risk of dementia late in life. People who exercise regularly in middle age are one-third as likely to get Alzheimer's disease in their seventies as those who do not exercise. Even people who begin exercising in their sixties can reduce their risk by as much as half. See you at the gym!

Specific changes in the brain's structure and function are associated with the deficits in memory and executive function during aging. The hippocampus becomes smaller with age, and this decrease in size correlates with memory loss. Similarly, the prefrontal cortex is important for working memory and executive function, and it becomes smaller with age as well.

Contrary to what you might imagine, brain shrinkage with aging is not due to the death of neurons. As you age, you do not lose neurons. Instead, individual neurons shrink. Dendrites retract in several regions of the brain, notably parts of the hippocampus and the prefrontal cortex. The number of synaptic connections between neurons in these areas decreases with age in most animals that have been examined. Older animals also have specific deficits in synaptic plasticity, the process that drives learning (see chapter 13), but only in certain parts of the brain.

On the other hand, some brain functions are not influenced much by age. Verbal knowledge and comprehension are maintained, and may even improve, as we get older. Vocabulary is another area that tends to be spared by aging. Professional skills are typically resilient, especially if you

Did you know? **I'm losing my memory. Do I have Alzheimer's disease?**

If you forget where you put your glasses, that's normal aging. If you forget that you wear glasses, then you probably have dementia. A disorder like Alzheimer's disease, which causes two-thirds of the cases of dementia, is not an extreme example of regular aging, but involves deterioration of specific brain regions along with symptoms that never occur in normal aging. People with advanced dementia cannot remember important incidents from their own lives and may not even recognize their spouses or children.

The strongest risk factor for Alzheimer's disease is simply age. The incidence of the disease doubles every five years after age sixty, reaching almost half the population by age ninety. Statistical estimates suggest that about 75 percent of people in the U.S. would develop Alzheimer's disease if we all lived to be one hundred. As the world's population ages, dementia is becoming more of a problem; its current incidence is twenty-four million people worldwide, and the number is expected to increase to eighty-one million by 2040.

Genetic factors have a considerable influence on your susceptibility to dementia, particularly the age of its onset. About a dozen genes have been identified as risk or protective factors, but one of them, the ApoE gene, has a stronger effect than all the rest put together. The average age of onset is about fifteen years earlier for people with two copies of the risky form of the ApoE gene compared to people with the protective form of the gene.

Many of the lifestyle factors that influence brain function during normal aging are also relevant to Alzheimer's disease. As discussed above, exercise is strongly protective. Other factors that correlate with a reduced chance of dementia include education, regular consumption of moderate amounts of red wine (but not beer or hard liquor), and the use of over-the-counter pain relievers with anticlotting effects, like aspirin and ibuprofen. In general, it seems as though improving your brain's ability to function tends to improve its resistance to a variety of problems, including dementia, late in life.

continue to practice them. Similarly, people who practice physical skills regularly are more likely to maintain them; in this case, there is some evidence that experts develop new strategies for well-rehearsed tasks to compensate for cognitive decline as they age. In general, anything that you learned thoroughly when you were younger is likely to be relatively spared by aging.

The good news is that older people have one important advantage over the young: a better ability to regulate their emotions. The frequency of negative emotions decreases with age until it levels off around age sixty, while positive emotions remain about the same. As people get older, they become less likely to perceive negative events or to remember those from their everyday lives or the past. Negative moods pass more quickly in older adults, and they are less likely to indulge in name-calling or other destructive behavior when they're upset.

There are also some more general changes in brain activity during aging. Older adults tend to activate more distinct brain areas than young adults while performing the same task. Compared with young adults, older people also tend to show lower overall brain activity and use areas on both sides of their brains instead of just one. These findings suggest that people use their brains differently as they age, even though they may perform a task equally well. This may be because older people learn to use new parts of their brains to compensate for problems elsewhere.

Cognitive decline at a certain age is not inevitable. Your lifestyle has a lot of influence on your abilities late in life. We mentioned before that people tend to retain skills and knowledge they learned thoroughly when they were younger. Perhaps for this reason, educated people have better cognitive performance with age than less-educated people. Another way to keep up your cognitive performance is to have intellectually challenging hobbies. This effect is more pronounced

Did you know? **Are you born with all the neurons you'll ever have?**

Many of us learned in school that the brain is unique because, unlike other organs of the body, it doesn't add new cells over its lifetime. Scientists believed this for many decades, but new discoveries indicate that it's not true. Both animal and human studies show that a few parts of the brain do produce new neurons in adulthood, though this ability declines with age. In particular, new neurons are born in the olfactory bulb, which processes smell information, and in the hippocampus. More of these new neurons survive and become functional parts of the brain's circuitry in animals that are learning or animals that exercise a lot. At present we don't have much information on the environmental conditions that encourage this process.

in blue-collar workers than in highly educated people, perhaps because educated people tend to work in jobs that involve considerable intellectual stimulation.

Attempts to improve cognitive skills in the elderly through training have yielded mixed results. Although most training programs work to some extent, the gains tend to be specific to the trained task, leading to little improvement in brain function across tasks. On the bright side, though, these gains can last for many years in some cases. One way to get around the problem of task specificity is to practice a variety of skills—either formally or by staying involved in several hobbies or volunteer projects in retirement. Our strongest suggestion, though, is to exercise consistently (see *Practical tip: How can you protect your brain as you get older?*), as keeping your heart in good shape has general positive effects on the brain, particularly on executive function, which helps you perform a variety of mental activities.

It seems that the Greeks were onto something when they recommended that people aim for a sound mind in a sound body. You'll be doing your best to keep your brain healthy if you keep some of both types of activity in your life. If you already get enough exercise, add an intellectual hobby like learning a new language or playing bridge. If you have an intellectual job, get a physical hobby like playing tennis or jogging. In general, having both physical and intellectual interests is the best protection against losing brain function with age.

CHAPTER 15

IS THE BRAIN STILL EVOLVING?

New technologies in transportation, medicine, electronics, communications, and weaponry have led to tremendous changes in our lives and habits over the last hundred years. Public health initiatives, vaccination, and sanitation have increased life expectancy by decades. Jet travel and communication have made the world a smaller place. Telecommunications and the Internet have made unprecedented amounts of information available to anyone, almost anywhere. Mass entertainment, with its constant stimulation, has become a major part of daily life. These advances have changed how we experience the world. Is the human brain also changing to keep pace?

Brains can change over time in two ways. First, the environment can influence brain development, leading to rapid changes, even within a generation. Second, there is biological evolution, which requires at least one generation to cause changes.

Rapid changes can be driven by the direct biological effects of a new environment. For instance, children growing up in preindustrial England faced challenges such as disease, nutritional deficiencies, and difficult field labor. After the Industrial Revolution, these were replaced by problems such as factory labor conditions, urban living, and pollution. Living conditions changed again and again through the Edwardian era, World War II, and the Cold War. Now children in developed countries grow up with standardized schooling, better nutrition, mass entertainment, computers, cell phones, and other technology.

Some of these changes in environment may underlie the Flynn effect, a phenomenon first noticed by New Zealand political scientist James R. Flynn. Using data from twenty countries from around the world, Flynn examined performance on standardized IQ tests over time. He

Did you know? **Understanding nature versus nurture**

What determines intelligence—genetics or your environment? The simple answer is both, but let's examine it a little. Genes have no effect without an environment, and vice versa. Both must interact during a child's development. The more interesting question is how they interact.

For many characteristics, your genes basically set an upper limit on your development. Take height, for instance. If we imagine two children with the same genes (like identical twins), the one who isn't fed enough protein while growing up (let's call him Tom) will end up shorter than the one who gets good nutrition (Mike). On the other hand, once Mike's basic nutritional needs have been met, stuffing him full of extra fish and chicken won't make him grow any taller, because he's hit his genetic limit. Instead, he'll just get fat. A third kid, Jeff, whose parents have passed along height genes with more potential but don't feed him as well, may end up the same height as Mike. Immigrants who move from poorer countries to richer ones often see their children grow much taller than they are. By the same process, economic development can increase the average height of a population.

Sam has seen this effect in his own family. He is six foot one, several inches taller than anyone in his parents' generation, all of whom grew up in prerevolutionary China. His brother Ed, at six foot six, towers over them all; his height is unheard of in the previous generation. As native-born Americans, they are examples of the height benefits that come from living in a highly developed country.

Intelligence works in a similar way, except that the environmental influences on its development are more complicated and less understood. Basic nutrition is important for any kind of growth, but brain development is probably also influenced by other factors, like social experience and intellectual stimulation. But by the same token, once the environment meets a certain standard of quality—albeit one that's not well defined—no amount of extra nutrients or stimulation will increase a child's intelligence beyond the natural limit imposed by genetics.

found that, within each country, the average scores were steadily higher for people who were born in later years—increasing about three IQ points per decade. In some nations, such as Denmark and Israel, IQ scores rose even faster, about twenty points over thirty years—little more than a single generation. For instance, in verbal and performance IQ, an average Danish twelve-year-old in 1982 beat the average scores of a fourteen-year-old from his parents' generation in 1952.

Changes in IQ over time imply that intelligence tests don't simply measure some pure, inborn capacity, but also track the effects of the environmental surroundings in which a person matures. Better nutrition and health can lead to better brain growth, and a more stimulating environment may also enhance brain development and function. Indeed, since we are highly social animals, these factors may be intensified by social interaction with other individuals who are also developmentally accelerated, leading to a positive feedback effect—and even more improved performance. Because of better nutrition and a more stimulating environment, it is entirely possible that people's brains today are, on average, more sophisticated than they were a hundred years ago.

Some evidence suggests that this effect has begun to level off. In Denmark, the nation with the largest past gains, IQ scores have stopped increasing in recent years. One possibility is that environmental effects can limit brain development, but only when resources are scarce (see *Did you know? Understanding nature versus nurture*). In other words, as the number of people who are poor or resource deprived decreases, the average IQ increases. This idea is supported by a recent study of Spanish children, which examined intelligence gains in the population over a thirty-year period. The IQ scores among the lowest-scoring children went up the most, with almost no gain in the top half of the population. Further support for this idea can be found in studies in the U.S., which show that, at poorer levels of society, educational achievement is correlated with the resources available in schools, but at richer levels, educational achievement is more strongly correlated with heredity and home environment.

However, all this progress does not mean that our brains are evolving. Instead, because the Flynn effect has occurred steadily over just a few decades, it cannot possibly be true evolution. Evolution usually refers to changes in genes that are passed on to offspring and would therefore require at least one round of reproduction and selection. This would lead to hereditary changes, so that a person born with the advantageous genes would eventually outperform other people brought up in the same environment.

An important thing to understand is that natural selection works through practical outcomes. It doesn't matter whether an animal knows how to find food because it's got an automatic program for food-finding tattooed on its brain at birth, or whether it is good at learning from its early experiences to get better at foraging. Either way, if that animal gets enough to eat, it will survive and be more likely to reproduce. For this reason, natural selection has produced brains that enable their owners to survive in the environment around them. Different animals may succeed by being adept at social interactions, or by being good at learning to survive in different environments. So nature versus nurture is the wrong debate; selection promotes genes that are especially good at getting along with their environments.

When people ask if the brain is still evolving, they often mean to ask whether the genetic

Did you know? **Machiavellian intelligence—a brain arms race?**

Primates are social—and mean. It's true for monkeys, and it's true for apes, including humans. We live in groups, compete with one another for food and mates, and are constantly forming and breaking alliances. The reasoning behind these social relationships can get quite convoluted, starting with "I like you; you like me" and ending up with "You pretend to like me when we are in front of her" and even "You and she might take my banana when I am not looking." It's a jungle out there.

Some have suggested that constant social competition is a main factor driving brain evolution in primates. Within the history of a species, social maneuvering over many generations may favor the selection of individuals with more mental firepower. This would lead to a brain arms race, in which increases in some animals' brain size would create pressure on other members in the species to keep up. Indeed, our species devotes more of its brain mass to cerebral cortex than any other species, about 76 percent. Chimpanzees are in second place at 72 percent, gorillas in third at 68 percent. Dolphins, although they have large brains in absolute terms, are considerably behind at 60 percent. In our case, the extra cortical volume turns out to be good for many things, like language and making tools.

Increased brain size could also open up new niches in the environment where a species may thrive. For example, though chimpanzees and gorillas are restricted to certain parts of Africa, humans were able to find a way through the geographic bottleneck leading from Africa to other parts of the world—and then adapt to a wide variety of conditions.

mechanisms that determine brain size or structure are changing. This is harder to answer because it can be many generations before any change at the evolutionary level becomes visible.

Human evolution by natural selection is hard to observe within a person's lifetime, but it is possible to study in animals with a short life cycle, so that many generations fit into a single lifetime of a human observer. For instance, in the Galápagos Islands, where food supply and weather vary strongly from season to season, finches with different beak types survive depending on the type and location of food available. Finches grow to adulthood and reproduce in just a few years. Over multiple generations, the range of beak types can change, moving toward long and narrow or short and stubby, depending on what is better for obtaining food. These changes have been seen in times as short as a single decade.

For natural selection to occur, individuals with a certain characteristic must have more offspring than individuals lacking that characteristic. Selection for differences in brain function is likely to be gradual; it may take millenia before any changes in intelligence become evident. Promoting the Flynn effect, which works much faster, is a better bet for improving our species—or at least a bet with a more immediate payoff.

If evolutionary change eventually does occur, however, it will be a continuation of processes already at work in the history of our species. There is evidence for relatively recent evolution of some of the genes that drive brain development ("recent" in evolutionary terms, meaning over the last ten thousand years). Two genes involved in brain development, *Microcephalin* and *ASPM*, have been studied in individuals around the world. These genes were originally discovered because they lead to severe defects in brain size or structure when missing or damaged. Persons with defective *Microcephalin* or *ASPM* are physically normal except that their brains are tiny; as a result, they are severely mentally retarded. This defect suggests that the proteins encoded by *Microcephalin* and *ASPM* are necessary in some way for normal development. This led to the speculation that the functionality of these proteins could also vary within the general population, and therefore lead to variation in brain size among individuals.

A team of researchers working with DNA from over a thousand people around the globe found that particular versions of *Microcephalin* and *ASPM* are inherited much more frequently than would be expected. This suggests that natural selection is at work. Based on comparisons with the rate of change in the rest of the genome over time, newer versions of the genes first appeared in the human population between six thousand and thirty-seven thousand years ago. The time is not known more precisely because DNA from that long ago has not been tested. Since generation times are typically fifteen to twenty years, these changes represent the cumulative outcome of hundreds to thousands of generations of selection.

It is also not known what the preferred versions of these genes are doing for people. So far, no correspondence has been found between gene version and brain size among normal humans, suggesting that brain size is determined by many additional factors. It is possible that these genes give some other advantage, such as a lower chance of developing a brain defect. The genes could even be involved in the development of other organs. Like the Flynn effect, defects in these genes may be a form of deprivation. In any case, the mechanisms that have driven increases in normal brain size are yet to be determined. Whatever these genes are doing, they fit into a larger story in which evolutionary genetic change in brain development takes thousands of years to accumulate. So don't hold your breath!

PART FOUR

YOUR EMOTIONAL BRAIN

CHAPTER 16

The Weather in Your Brain: Emotions

Most people assume that emotions interfere with our ability to make sensible choices—but that's not right. Emotions (unlike moods) occur in response to events in the world and keep our brains focused on critical information, from the threat of physical harm to social opportunities. Emotions motivate us to shape our behavior to gain what we desire and avoid what we fear.

Most real-life judgments cannot be based entirely on logic because the information we have is usually incomplete or ambiguous. It would be easy to decide whether to change careers if you could know in advance how well you would perform in the new job and how satisfying you'd find it. In most cases, though, you only have your intuition to go on. That's fine as long as your orbitofrontal cortex, a key part of the brain's emotional system, is working properly.

People with damage to this region have a rough time getting around in the world. One famous patient known as EVR was a financial officer with a small company and was happily married with two children when, at age thirty-five, he was diagnosed with a tumor in the front of his brain. Surgery to remove the tumor also took out a big piece of his orbitofrontal cortex. Afterward, he could still talk sensibly about the economy, foreign affairs, and current events, and reason his way through complicated financial and ethical problems. His memory and intelligence were unchanged, but he was not himself. He had trouble with even minor decisions, making lengthy comparisons between different shirts in the morning before grabbing one at random, for instance. More important choices eluded him as well. In short order, he lost his job, was divorced by his wife, and after entering into unwise business ventures that led to bankruptcy, eventually moved in with his parents. He married a prostitute and was divorced again after six months.

Such disastrous consequences are common among people with orbitofrontal damage (though the exact results of brain damage also depend on each individual's genes, life history, and personality before the damage). These patients remain able to plan and execute complex sequences of behaviors, but they do not seem to take into account the probable consequences of their actions. They do not show anticipatory anxiety before taking a big risk, and they are not embarrassed by socially inappropriate behavior that most of us would find mortifying. Indeed, they

Did you know? **Emotions and memory**

You probably remember more about your last vacation than about the last time you went to the post office. Psychologists have long known that emotionally intense events produce vivid memories. Emotional arousal seems to provide a particular advantage for the long-term storage of important details of an experience, sometimes at the cost of remembering peripheral details. People with damage to the amygdala do not show this enhanced memory of the central details of an emotional experience, suggesting that this brain region is important for the influence of emotion on memory. The amygdala appears to become involved in memory during intense situations, whether the emotions are positive or negative.

Emotional arousal causes the release of adrenaline, which activates the vagus nerve, part of the sympathetic nervous system (which controls the so-called fight-or-flight reflex). The vagus nerve projects to the brainstem, which then sends information to the amygdala and to the hippocampus, an area that is important for memory. An effect of this activity in both brain regions is to increase synaptic plasticity, a process that is thought to underlie learning (see chapter 13). Blocking the receptors for this information in the amygdala prevents adrenaline from enhancing memory, while activating these receptors in the amygdala improves memory.

Stressful situations also cause the release of glucocorticoids (stress hormones). These hormones act directly on the hippocampus and amygdala to enhance memory. Damage to the amygdala prevents the enhancement of memory by glucocorticoids in the hippocampus, suggesting that amygdala activity is necessary for this process.

Stress can also harm memory under some circumstances. Glucocorticoid hormones interfere with working memory by acting in the prefrontal cortex. Finally, chronic stress can damage the hippocampus (see chapter 10), leading to permanent memory deficits for all types of information, not just emotional memories.

don't seem to experience any of the social emotions under the appropriate circumstances, although they do experience emotions. This may be because they have difficulty monitoring their own behavior to determine how it relates to the rules of social interaction. When this damage is acquired in adulthood, patients can state these rules correctly but tend not to apply the rules to their own behavior. Those whose brains were damaged in childhood are unable even to describe the rules of social interaction, let alone apply them.

Now that we've explained why your emotional brain is important, let's look at its other parts. The amygdala is best known for its role in fear responses (see chapter 13), but it also responds rapidly to positive emotional stimuli. Overall, the amygdala seems to be important for focusing attention on emotionally salient events in the world. Neurons in the amygdala respond to sight, sound, or touch, and sometimes to all three. Many neurons have preferences for objects, especially for rewarding objects like food or faces. These preferences are modified by the animal's motivational state, so that a neuron that responds to fruit juice when the animal is thirsty stops responding once the animal has had its fill of juice.

Removal of the amygdala reduces some types of fear in animals and people. In particular, such damage reduces the physical signs of anxiety. When playing a card game, for instance, people with amygdala damage fail to respond to risks with increased heart rate and sweaty palms. (You might imagine that this would allow them to make a good living in Las Vegas, but that guess would be wrong. It turns out that this emotional reaction is necessary to allow people to make good decisions under uncertain circumstances.) Similarly, animals with amygdala damage respond less to anxiety-provoking situations, showing decreased vigilance and less freezing or flight.

Animals with damage to a particular part of the amygdala have difficulty with tasks that require revising the reward value of an object or situation, as might happen when you discover that the piece of chocolate that you just put into your mouth is actually licorice (no matter which one you prefer). These animals have normal preferences for tasty foods and work for rewards, but they lack the ability to adjust their preferences based on experience and can't learn to avoid foods that make them sick.

Most emotions are generated by a common set of brain regions, but there are a few emotion-specific regions. Certain types of brain damage can impair the experience of disgust or fear without affecting other emotional reactions. We will examine the amygdala's role in fear more closely in chapter 17.

Disgust is evolutionarily old, dating back to the need of foraging animals to determine whether a food is good to eat. The key brain regions for generating feelings of disgust are the basal ganglia and the insula. Electrical stimulation of the insula in humans produces sensations of nausea and unpleasant tastes. Rats with damage to either of these areas have difficulty learning to avoid foods that make them sick; in people, the role of these regions has broadened to include recognizing similar feelings in others. Patients with damage to these regions have difficulty recognizing facial expressions of disgust, as do people with Huntington's disease, a primarily motor disorder, which is caused by degeneration of neurons in the striatum (part of the basal ganglia).

Remarkably, these same brain regions seem to cause us to wrinkle our noses not only at spoiled food but at violations of moral decency. For instance, the insula is active when people think about experiences that make them feel guilt, an emotion that has been described as disgust directed toward oneself.

More generally, the insula's job seems to be to sense the state of your body and trigger emotions that will motivate you to do what your body needs. You can't always trust what your body thinks it needs, of course, and the insula has also been implicated in cravings for nicotine and other drugs. The insula sends information to areas involved in decision making, such as the anterior cingulate and prefrontal cortex. The insula is also important in regulating social behavior: it helps us infer emotional states (such as embarrassment) from physical ones (such as a flushed face). The insula is one of several brain systems that responds in a similar way both to one's own action or state and that of another person; another is the mirror neuron system (see chapter 24).

We share emotions—and the brain systems that produce them—with other animals. However, human emotions are particularly complex, in part because we have such a large frontal cortex. Though mice can be frightened, it's hard to imagine a mouse feeling ashamed. Emotions control many of our social behaviors, so it should come as no surprise that the brain regions that are important for emotions are also important for processing social signals. So-called social emotions, such as guilt, shame, jealousy, embarrassment, and pride, arise later in development than the basic emotions of happiness, fear, sadness, disgust, and anger. These emotions guide our complex social behavior, including the desire to help other people and the urge to punish cheaters, even at a cost to ourselves. Brain imaging experiments show that people with stronger activity in emotional brain areas in response to such situations are more likely to be willing to pay the cost of altruism or enforcement of social norms.

How we think about a situation often influences our emotional reaction to it. For example, if your date failed to show up at the restaurant on time, you might be angry that he'd been so inconsiderate of your feelings, or you might be afraid that he'd been in a car accident. When you later learned that he had been delayed because he'd stopped to help someone who'd had a heart attack, you might feel happy and proud.

These situations show how our brains can modify our experience of emotions based on our intentions or on how we perceive events. Several areas of the cortex send information to the core emotion system to modify our perception of an emotional response. The simplest form of emotion regulation is distraction, turning your attention to something else, usually temporarily. When distraction is working, functional imaging studies show that the activity in emotional brain areas is decreased. Distraction can decrease the negative emotions associated with physical pain, in part by reducing activity in some pain-responsive areas like the insula while increasing activity in areas associated with the cognitive control of emotions, mainly in the prefrontal and anterior cingulate cortex. Similarly, anticipating an experience that is likely to produce either positive or negative emotions can often activate the same brain regions that would normally respond during such an experience.

A distraction-like effect can also be brought under conscious control. For instance, some yoga masters claim not to feel pain during meditation. When one of these masters was put in a brain scanner and asked to meditate, a laser beam stimulus that would normally be extremely painful caused no sensation—and led to very little response in the insula.

A more lasting way to regulate your emotions is called reappraisal. That's when you reconsider the meaning of an event as a way of changing your feelings about it. For example, if your toddler touched a hot stove and burned her hand, you might initially feel angry that she disobeyed you and guilty that you weren't attentive enough to stop her from getting hurt. On further reflection, though, you might realize that the injury was not very serious and would heal quickly, and that your daughter had learned a valuable lesson about the importance of listening to your instructions. Both those interpretations could make you feel less upset about what had happened.

Did you know? How does your brain know a joke is funny?

Humor is hard to define, but we know it when we see it. One theory suggests that humor consists of a surprise—we don't end up where we thought we were going—followed by a reinterpretation of what came earlier to make it fit the new perspective. To make it a joke instead of a logic puzzle, the result needs to be a coherent story that isn't strictly sensible in everyday terms. Some patients with damage to the frontal lobe of the brain, particularly on the right side, don't get jokes at all. Typically this is because they have trouble with the reinterpretation stage of the process. For instance, given a joke with a choice of punchlines, they can't tell which one would be funny. Laughter or feelings of amusement have been evoked in epileptic patients by stimulation of the prefrontal cortex or the lower part of temporal cortex. Functional imaging studies show that the orbital and medial prefrontal cortex are active when people get a joke. Since humor includes both emotional and cognitive components, it makes sense that these prefrontal regions, which integrate the two functions, would be involved.

Humor also makes people feel good, apparently by activating the brain reward areas that respond to other pleasures like food and sex, as we discuss in chapter 18. Especially when coupled with surprise, a sense of pleasure can trigger laughter. Indeed, laughter may be an ancestral signal that a situation that seems dangerous is actually safe. Multiple types of humor activate areas that respond to emotional stimuli, like the amygdala, midbrain, anterior cingulate cortex, and insular cortex. The last two regions are also active in situations of uncertainty or incongruity, so they may participate in the reinterpretation stage of getting a joke. The funnier a person thinks a joke is, the more active these areas (and the reward regions) are.

Humor's rewards go beyond simply feeling good. Being talented at making other people laugh can improve all sorts of social interactions, helping you to find a mate or communicate your ideas effectively. Humor also reduces the effects of stress on the heart, immune system, and hormones. So if you're the kind of person who tends to be amused by things that other people don't find funny, remember that you're likely to get the last laugh.

Reappraisal seems to rely on the prefrontal cortex and anterior cingulate cortex. In imaging studies, people attempting to reinterpret emotional stimuli show activation of these regions. Successful reappraisal results in changes in other emotion-related brain areas that are consistent with outward emotional changes, such as a decrease in amygdala activity when someone reappraises a stimulus to make it seem less scary. These brain changes are strikingly similar to the activity patterns in response to a placebo drug, another example of how people can experience an identical situation in different ways depending on their individual beliefs.

People who are good at reappraisal tend to be emotionally stable and resilient. Many of the gains that people make in psychotherapy can probably be attributed to improvements in their ability to reappraise situations in productive ways. In general, as mammals with a big frontal cortex, we are in a good position to train our emotional responses. Reappraisal, unlike most mental capacities, improves with age, perhaps as a consequence of maturation of the prefrontal cortex, or maybe just from practice. This may explain why mature adults tend to be happier and experience fewer negative emotions than young adults.

So the next time someone says, "Don't be so emotional," you'll know better. Your emotions—both pleasant and unpleasant—provide a sensitive guide to effective behavior, helping you to predict the likely consequences of your actions when you don't have enough information to decide logically. Go ahead and be emotional. As long as your emotion regulation system is in good working order, it's likely to be the right choice.

Humor can be dissected as a frog can, but the thing dies in the process and the innards are discouraging to any but the pure scientific mind.

—E. B. White

CHAPTER 17

DID I PACK EVERYTHING? ANXIETY

We're not trying to make you nervous, but the truth is that being too relaxed can kill you. In a world filled with hazards, worrying can offer big advantages for survival. Of course it is possible to worry too much—for instance, if you're a badger who's too fearful to leave his den to find food or a mate. It's also possible to worry about the wrong things, as when a person develops a phobia that turns entering a dinner party into a heart-poundingly scary experience. On the whole, though, anxiety serves many useful purposes, and not just in leading us to exercise caution in the face of danger. Anxiety also motivates positive behaviors, from finishing an assignment before the deadline to storing enough food to get through the winter. Ironically, emotions that make us feel bad often cause us to behave in ways that are good for us, which is why they have become so common.

Although everyone experiences anxiety sometimes, people (and other animals) show individual differences in how easily their anxiety is triggered, how intense it is, and how long it lasts. Some of these individual differences are due to our genes. Having a relative with panic disorder (discussed below), for example, increases your risk of developing the disorder by a factor of about five.

Genes not only control baseline anxiety levels but can also determine our sensitivity to life stressors, such as child abuse, the death of a parent, or divorce. People with the protective variant of a particular gene, for instance, can handle a lot of tough events with little chance of getting an anxiety disorder or depression as a consequence. This gene encodes the serotonin transporter, which removes the neurotransmitter serotonin from the synapse after it has done its job. People with the vulnerable variant of the gene are more sensitive to stress, but they can get

Myth: The car-crash effect

People often report that during a sudden dangerous event, such as an automobile crash, time seems to slow down. Afterward, they say they were able to evaluate the situation, consider alternatives, and take evasive action in a matter of moments. Such an ability would clearly confer a tremendous survival advantage.

In a sense, time does slow down under stress—or, more accurately, people perceive it to slow down. To test performance speed during fear, researchers used a very exciting but harmless scenario, an amusement park ride. The ride in question is a free-fall experience in which helmeted participants are dropped one hundred feet into a waiting net.

To measure perceptual speed during the fall, the researchers mounted a small video monitor on the wrists of participants. On the screen was a sequence of rapidly changing images of a letter or number (for instance, a black 1 against a white background) alternating rapidly with a canceling image (a white 1 against a black background). They sped up the images just enough so that under normal conditions, participants saw only a uniformly gray screen. Then they dropped the participants from the edge, instructing them to keep their eyes on the monitor.

The falling participants did not perceive the digits with any better accuracy than participants who performed the same task with their two feet planted firmly on the ground. Thus, temporal perception did not improve, even though participants believed the fall to last much longer than it actually did. In separate measures, participants estimated their own fall to last 36 percent longer than others' falls.

These results indicate that there are separate mechanisms underlying duration judgments and temporal resolution. Even though you might think that an event took a long time, you cannot become like Neo in *The Matrix*, seeing the world in "bullet time." In dangerous situations, one possibility is that neurotransmitters, such as adrenaline, cause memories to be laid down more richly in a given period of time without a speed-up in sensory processing. A remaining question is how to measure whether mental processing is faster during very exciting moments. Bungee-jumping and Sudoku, anyone?

along fine if nothing too bad happens in their lives. People with one copy of each variant (because we all have two copies of every gene, as you may recall from high school science) fall somewhere in the middle. They can handle one bad event, but multiple bad events may send them over the edge into depression or an anxiety disorder.

The only thing we have to fear is fear itself.

— Franklin D. Roosevelt

Anxiety disorders are the most common type of psychiatric disorder in the U.S., affecting about forty million people. As many as 90 percent of people with anxiety disorders also have clinical depression at some point in their lives, and many of the same treatments are effective for both problems. For example, selective serotonin reuptake inhibitors like Prozac, which are commonly used to treat depression, also work well for anxiety disorders. This overlap suggests that the brain mechanisms that cause depression and anxiety may be similar, though the origin of abnormal anxiety is better understood.

As we have already said (see chapters 13 and 16), damaging the amygdala interferes with fear responses and fear learning in humans and other animals. Stimulating the amygdala produces fear responses in animals. You don't need a brain scanner to tell you when your amygdala is active: it's happening when your heart races and your palms get sweaty. Your blood pressure also goes up, and, in extreme cases, you may find it hard to breathe. These symptoms occur because the amygdala has a direct connection to the hypothalamus, which controls the body's stress responses. Amygdala activity leads to activation of the sympathetic nervous system (the fight-or-flight response) and release of glucocorticoid stress hormones. People who experience intense and acute onset of these symptoms are said to have panic attacks, a type of anxiety disorder that can produce symptoms so overwhelming that people believe they're about to die.

An overactive amygdala probably causes some anxiety disorders. Other patients seem to have normal amygdala responses. Instead, they have a problem with the prefrontal cortex, which is responsible for turning off anxiety when it's not appropriate for the situation. The amygdala receives input directly from the senses, so its responses are designed to be fast, not accurate. Often, further analysis by a more careful part of the brain leads to the realization that there's nothing to fear. (You thought you saw a snake, but it turns out to be a branch swaying in the breeze.) The prefrontal cortex then inhibits the amygdala, shutting down the anxiety. If this process isn't working correctly, people will continue to feel anxious long after the danger has passed. Some of the best treatments for anxiety disorders probably work by increasing the effectiveness of this inhibitory pathway.

Mild anxiety shouldn't require professional treatment. If you want to try the self-help approach, start by thinking about how to reduce stress in your life. You can do this in two main ways: reduce your exposure to stressful situations or learn better skills for coping with them.

Did you know? Post-traumatic stress disorder

Some rape victims, combat veterans, and others who've experienced extremely traumatic events develop post-traumatic stress disorder (PTSD). People with this disorder are constantly on guard, which leads them to be easily startled and to have difficulty sleeping. They also relive the traumatic events during nightmares or intrusive daytime thoughts, and they may become emotionally detached and lose interest in everyday activities. PTSD symptoms persist throughout life for about 30 percent of sufferers. PTSD is not a modern invention. Its symptoms were described in ancient times, a famous example being the transformation of Achilles by war in the *Iliad*. Indeed, PTSD has occurred in all wars that have been studied.

Most adults have experienced at least one traumatic event of the type that can cause PTSD, though only some people develop the disorder after a trauma. The strongest trigger is trauma deliberately caused by another person, such as rape or kidnapping. About half of rape victims go on to develop PTSD, while natural disaster victims have a relatively low risk (about 4 percent). The same treatments are helpful for PTSD as for other anxiety disorders, but progress can be much slower. Persistent PTSD has negative consequences for the patient's work and relationships that tend to linger after the anxiety itself starts to fade.

Like other anxiety disorders, PTSD is twice as common in women as in men. (In the U.S., women have a 10 percent chance of developing the disorder in a lifetime, while men have only a 5 percent chance.) There are two proposed explanations for this difference. One is that women experience more traumatic events (or more intense trauma), as rape and spousal abuse are substantially more common for women, though men certainly experience more combat-related trauma. The other is that women are more sensitive to fear learning or stress, which may make them more vulnerable to anxiety disorders. The evidence for this idea is weak and inconsistent, but it is true that more women (20 percent) than men (8 percent) develop PTSD after a traumatic event. Of course, it's possible that both these explanations may contribute to the gender disparity.

People with PTSD also show reductions in hippocampus size compared to people without the disorder. At first, scientists thought that this happened because PTSD causes stress, which is known to damage the hippocampus. It turns out, though, that when researchers looked at identical twin pairs, in which only one twin had combat experience, a small hippocampus in the twin who stayed home from the war was a good predictor of whether the other twin would get PTSD in combat. This finding suggests that certain people are predisposed to get PTSD, perhaps because their brains are hyperresponsive to stress.

Which of these approaches is most useful will depend on what's causing your stress. One good way of living more comfortably with stress is to exercise regularly, preferably at least thirty minutes every day. Exercise improves mood and, as we learned in chapter 14, has the added benefit of helping to preserve brain function and reduce the risk of dementia as you age, so there's really no downside. Meditation may also reduce stress responses. Some people find yoga particularly helpful, as it combines exercise with mental calming. You should also try to reduce your caffeine intake and get enough sleep. Resist the temptation to medicate anxiety with tranquilizers or alcohol, which will only make the problems worse in the long run; many people with anxiety also suffer from substance-abuse disorders. If these techniques don't reduce your anxiety or if your anxiety causes serious problems in your life, you may need to see a professional therapist.

Two types of psychotherapy, which are often used together, have proven to be effective for anxiety disorders in clinical studies. Both approaches are short-term interventions that concentrate on teaching patients to control situations that make them anxious, and both require active participation from patients. Behavioral therapy is based on extinction learning, which you may recall from chapter 13. Repeated exposure to a feared object or situation without negative consequences results in extinction, a process that teaches the animal or the patient not to fear the stimulus. Behavioral therapy focuses on helping people to stop avoiding anxiety-provoking situations, so they can learn that these situations are not really dangerous (see *Practical tip: How to treat a phobia*). Cognitive therapy focuses on helping people to learn how their thought patterns contribute to their discomfort and to substitute more productive ways of thinking about the problem, by distinguishing between realistic and unrealistic thoughts, for instance.

Practical tip: **How to treat a phobia**

A phobia is an intense fear of something that is not really that dangerous. People may develop irrational fears of anything from spiders to heights to social interactions. Phobias commonly start in childhood or adolescence, suggesting that they may be learned, but most patients don't remember a specific incident that triggered their fear. The tendency to acquire phobias seems to be partly due to genetic factors.

The good news is that phobias are among the most treatable of psychiatric disorders. Short-term behavioral therapy that focuses on desensitizing the patient's fear is very effective. Sometimes this approach is supplemented with drugs to temporarily reduce the fear and make it easier to face, or with cognitive behavioral therapy to encourage the patient to rethink her attitude toward the fear-inducing stimulus.

The therapist slowly exposes the patient to the feared situation in small steps, consulting often with the patient to be sure that the anxiety stays within a tolerable range. For instance, for a phobia of heights, the patient might first look at a picture taken from the second floor. Then the patient might imagine standing on a balcony and eventually an even higher place. As the anxiety fades, the patient would be exposed to real anxiety-producing situations in a controlled way to demonstrate that they are not really dangerous. This approach, in the hands of a trained therapist, has a good track record of bringing phobias under control.

Before you see a therapist, you should make sure you know what type of therapy he or she practices and whether it works for the problem that you want to have treated.

Doctors are testing some exciting variants to these approaches for treating anxiety disorders, though these new treatments are not yet widely available. Because the demand for behavioral therapy exceeds the number of trained therapists, some researchers are working on computer systems that allow people to control their own exposure to anxiety-producing situations. Another approach is to expose patients to a simulated version of the situation. Doctors have used virtual reality therapy to treat phobias, panic disorder, and PTSD. Preliminary evidence suggests that it may be as effective as direct exposure to the fear-triggering stimulus. In one particularly exciting new approach, doctors asked patients to take a drug called D-cycloserine before virtual reality behavioral therapy sessions. This drug activates NMDA receptors, which are important for learning. By improving learning, the drug increases the rate of fear-extinction learning during behavioral therapy. Patients in this study showed reductions in anxiety after as little as two sessions, and the improvement lasted for three months.

This group is now testing the same approach for treating PTSD in veterans of the Iraq war, who have an 18 to 30 percent risk of the disorder. If these treatments live up to their promise in further testing, it might be possible to greatly decrease the number of people who struggle with excessive anxiety.

Of course, we can't expect to eliminate anxiety altogether by using any of these techniques. If that happened, we would never get anything done. There's definitely an optimal level of anxiety—not so low that you lie on the couch all day, but not so high that you huddle under the bed—and unfortunately the best level for survival isn't necessarily the one that makes us feel most comfortable. But if anxiety is interfering with your life, we strongly encourage you to do something about it. Don't let a problem with anxiety take control of your life.

CHAPTER 18

Happiness and How We Find It

Timothy Leary would have been disappointed to learn that some of the happiest people in the U.S. are married, churchgoing Republicans who make more money than their neighbors. He might have been more pleased, however, to know that happy people also have a lot of sex and socialize frequently.

People's happiness tends to be determined by comparison to other people. Average income in the U.S. has risen steadily over the past fifty years, but the percentage of people who consider themselves very happy has stayed about the same, presumably because the standard for comparison has risen along with the average income. Thus, the important determinant of happiness is not absolute wealth but relative wealth—as long as you make enough that your basic needs are secure (about $30,000 per year). This means that most of us would feel happier to make $50,000 a year in a job where the local average salary is $40,000 than to make $60,000 where the average salary is $70,000. The things we could buy with the extra $10,000 each year wouldn't come close to compensating us for the happiness that we would derive from being paid better than our coworkers.

As one researcher says, "The key to happiness is low expectations." When you're making a major purchase, it's worthwhile to remember that ultimately you won't be comparing your new acquisition to the other possibilities in the store, but instead to what you already own—or what your friends own. Indeed, people tend to be less satisfied with their decisions when they have to choose among many options than when only a few options are available, suggesting that making more comparisons may reduce happiness by causing us to regret the options that we were unable to choose.

Even major life events have less lasting influence on happiness than you might guess. For example, blind people are no less happy than people who can see. Married people are, on average, happier than unmarried people (see *Did you know? How scientists measure happiness*), but having children has no overall effect on happiness. It seems that after a strong transient response to most good or bad events, people's happiness tends to return toward their individual "set point," which is mildly positive on average. This is called adaptation, and it's the reason that some people keep buying stuff they don't need: if having something new makes you happy, you have to keep renewing the feeling by buying more stuff because the effect never lasts.

Did you know? **Happiness around the world**

In the U.S., happiness differences between individuals don't depend strongly on demographic factors like income, but things change when we compare across countries. The explanation may be that because of the relative amount of wealth and stability in this country, happiness differences among Americans based on economic and political circumstances are not significant. On the other hand, the nations of Africa and the former Soviet Union contain some of the unhappiest people in the world, presumably due to widespread poverty, poor health, and political upheaval. Researchers from the Economist Intelligence Unit reported that 82 percent of the differences in average happiness between countries can be predicted from nine objective characteristics. Starting from the most important, these characteristics were health (life expectancy at birth), wealth (gross domestic product per person), political stability, divorce rate, community life, climate (warmer is better), unemployment rate, political freedom, and gender equality (the more even the ratio of male to female income, the happier the people).

Cultural factors also seem to affect happiness. For example, people in Denmark consistently report substantially higher levels of happiness than people in Finland, although the countries are similar on most demographic variables. A Danish research group provided a tongue-in-cheek explanation for this difference: on the same survey, Danes report having lower expectations for the upcoming year than Finns.

Did you know? How scientists measure happiness

If the idea of studying happiness sounds too touchy-feely for you to take seriously, you're not alone. There are some real limitations to this sort of research, but it's more reliable than you might think. The usual method for collecting data in such studies is pretty simple: researchers call up and ask people how happy they are. ("How satisfied are you with your life as a whole these days? Are you very satisfied, pretty satisfied, not very satisfied, or not at all satisfied?") Then they ask about a bunch of other stuff like people's income, marital status, and hobbies. When they have this information from a significant sample (typically thousands of people), they try to figure out what kinds of answers are more likely to come from happy people than from unhappy people.

This approach to research is called correlational, and it does have one big drawback. If you find out that two things routinely occur together, then it's a pretty good bet (though not guaranteed) that there's some relationship between them—but you still can't tell what the relationship is. For example, knowing that married people are happier on average than single people doesn't tell us whether your son would be happier if he got married, no matter what you might personally believe. Being married could make people happier, or being happy may simply make it easier to get married. Indeed, psychologists who measure happiness in the same individuals through several years of their lives have found that both these statements are true. Happier people are more likely to get married; that, in turn, makes them happier still. Not all happiness research is correlational, but when we interpret studies of this type, we need to remember that a correlation between two things can't tell us what we would most want to know: which thing causes the other, or if there is a third, unknown cause of both things.

Another point to keep in mind is that, as with most psychological research, the answer you get depends greatly on how you ask the question. For instance, when women were asked to list the activities that they particularly enjoyed overall, "spending time with my kids" topped the list. In contrast, when other researchers asked women to describe how they felt during each of their activities the previous day, the average positive rating given to interacting with children indicated that this activity is roughly as rewarding as doing housework or answering e-mail. This finding suggests that women find their children more rewarding in theory than in practice, at least on a moment-to-moment basis.

In its strongest form, the adaptation idea suggests that all efforts to increase happiness in an individual or society are futile and that people's life circumstances have no long-term influence on their happiness. This would be pretty surprising and almost certainly isn't correct. Indeed, some circumstances are reliably associated with unhappiness, including chronic pain or having to commute a long way to work.

The life events most likely to have a lasting negative influence on people's happiness include the death of a spouse, divorce, disability, and unemployment. In all these circumstances, people still adapt—their happiness is much more strongly affected right after the event and then moves back toward the baseline—but the adaptation is not complete. Even eight years after the death of a spouse, surviving partners remain less happy than they were when their spouse was alive. Deliberate attempts to increase happiness have also had some lasting success, though these interventions seem to be most effective if they are repeated frequently (see *Practical tip: How to increase your happiness*).

When psychologists follow the same people over time, most of them report fairly stable happiness. In one study of Germans over a seventeen-year period, the happiness of only 24 percent of the participants changed significantly from the start to the end of the study, and only 9 percent changed a lot. All individual circumstances—marriage, health, income, and so on—taken together account for only 20 percent of the differences in happiness from one individual to another in the U.S., while genetic factors account for about 50 percent of the differences. Identical twins reared apart (usually because they were adopted separately) are much more similar to each other in their adult happiness than fraternal twins who are reared apart, and about as similar in happiness as identical twins reared together. (The mysterious remaining 30 percent includes measurement errors, such as the differences between individuals in defining survey responses like "mostly satisfied.")

In general, the brain seems to respond more strongly to changes than to persistent conditions, right down to the level of single cells. Neurons also show adaptation (though they typically do it in less than a second, not months). Adaptation is efficient because most of the information in the world is stable, while most of the action that is important to your brain lies in the part of the world that is changing—objects that are moving, your mate's new facial expression, or an unexpected source of food. If the brain can cheat by devoting its limited resources to representing the information that is new, it may be able to more effectively help you respond to the world.

Neurons in several brain areas respond specifically to events that are "rewarding." A reward makes you more likely to repeat the behavior that led to the reward; examples are food, water, sex, and a variety of more complicated things like positive social interactions. In people, we

Practical tip: **How to increase your happiness**

Happiness is a moving target. Because of adaptation, frequent small positive events have a greater cumulative impact than occasional large positive events. Similarly, the elimination of daily irritants like commuting is likely to provide a substantial improvement in happiness. It's hard to believe that it would make you happier to spend fifteen minutes every evening for the rest of your life having a relaxed drink with a sympathetic friend than it would to win the lottery, but it's almost certainly true.

What makes people happy day to day? Women who were asked to recall their emotions at the end of each day rated having sex as the most rewarding activity, considerably ahead of the runner-up, socializing with friends. Indeed, more sex correlates with more happiness—and unlike money, the happiness-producing effects of sex do not diminish once you have enough of it. How well people slept the previous night has a stronger correlation with their enjoyment of the day than their household income. Setting realistic goals and achieving them is also associated with happiness for most people. You probably don't need to worry too much about varying your daily routine, as people who stick with their old favorites are happier than people who seek variety for its own sake.

The study of happiness is still in its infancy, but a few researchers have shown that behavioral exercises can increase happiness. The exercises are most effective if you do them consistently. Here are a few of the exercises that work:

- **Focus on positive events.** Every evening for a month, write down three good things that happened that day and explain what caused each of them. This exercise increased happiness and reduced symptoms of mild depression within a few weeks, and the effects lasted for six months, with particularly good outcomes for people who continued to do the exercise.

- **Practice using your character strengths.** You can find out what your strengths are by going to http://www.authentichappiness.org and taking the VIA Signature Strengths questionnaire. (The Web site is run by Martin Seligman, a well-known positive psychologist at the University of Pennsylvania. You'll need to register for access to the site, but the tests are free.) Once you know your top five strengths, make a point of using one of them in a new way every day for a week. This exercise and the previous one grew out of Seligman's research, as described in his book, *Authentic Happiness*.

- **Remember to be grateful.** Every day write down five things that you are thankful for. People who did this exercise for several weeks had more positive feelings and fewer negative feelings than people who did a placebo exercise. However, we do not know whether the effects are long-lasting, as the subjects were only followed for a month.

know rewards are associated with a subjective sense of pleasure, and people, like other animals, are willing to work for them (as well as for human-specific rewards like money). However, opportunities to record the responses of individual neurons in humans are rare, so studies of this sort are typically done with rodents or monkeys.

Scientists can distinguish between neurons that respond to rewards and neurons that respond to other aspects of a stimulus, like taste. The reward neurons are those that stop responding when the animal no longer wants the reward, like when a rat is no longer interested in a food because it has had enough to eat (though presumably the food still tastes the same). These neurons are found in brain regions like the orbitofrontal cortex, striatum, and amygdala, and they often respond not only to the existence of a rewarding stimulus, but also to some particular characteristic of the reward. For instance, one neuron might respond to one type of food but not another, or to a small reward but not a large reward. Although different neurons within a given brain area have different preferences, the same set of brain areas is active when the animal receives a lot of different rewards, from food to sex to the opportunity to spend time with its mate.

Some such neurons release the neurotransmitter dopamine. These neurons are located in the substantia nigra and the ventral tegmental area of the midbrain, and they project their axons to a variety of other brain regions that contain reward-responsive neurons, including those discussed above. These neurons seem to be specifically involved in reward prediction. Dopamine neurons are activated by unexpected rewards. For instance, experimenters taught rats that they could press a lever and get a reward—but only after a light was on. During the early stages of training, neurons were active when the food arrived. Later on, after the animals knew the task, the dopamine neurons began to fire as soon as the light went on—when the animal first knew it was going to get some food—and they were inhibited when the food failed to show up on schedule. When enough disappointments happened repeatedly, the neurons stopped firing in response to the light, and the animals stopped pressing the lever. In a variety of situations, then, these neurons appear to tell animals about which features of the environment predict when they will receive a reward.

What do dopamine or reward-responsive neurons have to do with happiness? We don't know how to define happiness in rats (it's hard enough to define in people), but it does look as though dopamine helps rats—and people—to choose behaviors that lead to positive outcomes. Evidence that signaling reward is one of dopamine's functions in people comes from Parkinson's disease, a movement disorder that involves the progressive death of dopamine-making neurons serving multiple functions. In addition to their motor problems, Parkinson's patients have difficulty learning through trial and error. When medication makes their dopamine levels high, Parkinson's patients learn more about responses that are paired with rewards. In contrast, when patients are not taking medication, and their dopamine levels are low, they learn more easily about responses that are paired with negative consequences. These results suggest that dopamine is involved in learning to choose behaviors that lead to positive outcomes, which sounds like a key ingredient in happiness to us.

Success is getting what you want. Happiness is wanting what you get.
—Unknown

CHAPTER 19

WHAT'S IT LIKE IN THERE? PERSONALITY

It's never pleasant to be disliked by someone you work with, especially when that someone plays anonymous practical jokes. However, as Shelley found, it can be a little less unpleasant when that someone is six inches long and has no bones—indeed, has no hard parts except for a beak.

Shelley spent one summer at the Marine Biological Laboratory, a research center on Cape Cod, Massachusetts, working with cuttlefish. Cuttlefish are members of the cephalopod family, a strange group of big-brained, big-eyed, multilimbed marine animals; their close relatives include octopuses and squid. That summer, Shelley spent her days in a small room with a cuttlefish in a tank next to her while she prepared behavioral tests for the animal. One day she felt something wet on her backside. She turned around, and saw nothing—just a cuttlefish in the tank. She assumed it was just a random splash from the aquarium pump. As it turns out, it was a pump—just not a mechanical one. She was splashed again several times before she realized that the water was coming from the cuttlefish itself. All cuttlefish have a siphon that they use to send water in specific directions. This particular cuttlefish was using its siphon on Shelley, but only when her back was turned. Somehow, it's hard to shake the sense that Shelley was the victim of a repeated expression of dislike by her crotchety experimental subject.

It's clear that individual animals have distinctive personalities, and that personality is at least partially inherited. Dog fanciers will gladly explain in great detail the quirks of different breeds. Pomeranians are high-strung; pugs, agreeable and unaggressive. One can see the whole range of behavior on display on any sunny day in a dog park. Personality also varies among species: we present exhibit A, the notable absence of cat parks.

Most of our interest in animal personality stems from our encounters with companion animals, such as dogs and cats. But ethologists (scientists who study animal behavior) examine individual personality and temperament in many species, from dairy goats and horses to guppies and spiders. They find that individuality appears to be a biological imperative that may be essential to survival strategies for any species. The research has illuminated what personality is good for and how it can be shaped by inheritance, development, and experience. It has even produced some early glimmerings of how brain mechanisms generate animal—and human—personality.

Many traditional psychologists have avoided the study of differences among animals altogether. A pioneer in animal behavior research, B. F. Skinner made a point of giving tests to animals under conditions that made the responses as reliable as possible. He designed his famous Skinner box to remove any distracting stimuli that might lead to environmental variation. Skinner's idea of a perfect experiment was one in which there was no variation from individual to individual; in principle, if you had a good experiment with one animal, you might use a second one only to make sure everything was okay.

There is a plausible reason for discounting talk of personality, not only in animals, but even in people. We constantly map our own actions to our individual motives and preferences, and we tend to assign similar motives and preferences to the actions of others. But this is a slippery slope. As pointed out in chapter 1, your brain is constantly lying to you about your own reasons for acting. We inadvertently create mental models of how things work expressed in terms of agency, even when the things in question are inanimate objects. For instance, it is common to describe a car as temperamental, or a house as personal and inviting. Yet nobody would attribute literal personality to these objects.

Ethologists wrestle with this problem continually. Their answer is to work with behaviors that are directly observable. Did the animal attack? Did it retreat? Did it curl up in a corner? In some sense, Skinner was just as focused on observing behaviors that were quantifiable. But ethologists' interest in differences has enabled them to catalog individual traits and try to understand the reasons for them.

One very striking finding is that not only do animals have individual personalities, but individuals can be categorized according to the same groupings and qualities that we use in classifying human personality. In one pioneering study at the Seattle Aquarium, researchers were able to break down octopus temperament into three principal dimensions: activity, reactivity, and avoidance. These measures did a good job of describing how the animals would react in various situations that included a human observer sticking her face into the tank, waving a test tube cleaning brush near the animal, or dropping a tasty crab into the water. Over time, and in controlled situations, researchers were able to reliably predict the tendency of individual octopuses to attack, retreat, or remain calm.

The variability in temperament in octopuses and many other species raises the question of what evolutionary sense it makes for a species as a whole if the natural behavioral tendencies—the temperament—of individuals vary. One possibility is that different personalities can adapt to different niches in the environment. For instance, being daring might get an animal to the front of the line in grabbing the food, but if many dangerous predators are nearby, that animal would also be at greater risk of being eaten. In such a situation, the stay-at-homes could lie low, then come out to cadge a few scraps—and live to see another day. Likewise, extroverted people may get more dates, but they also have more accidents and therefore end up in the hospital more often. Finally, consider an extreme example, the female North American fishing spider. Some fishing spiders are extremely aggressive hunters and are always first to grab passing food. But these same females have trouble during the mating season, when they can't keep their legs off of their suitors—and then eat the poor guys before they have the chance to mate. Oops.

Variation may be a strategy that helps a species to survive in an ever-changing world. The world changes much faster than species do as a whole, as adaptation through genetic change takes many generations. Sexual reproduction rescues us from this plight. Each individual of a species takes DNA from his or her parents and combines it to make a new, genetically unique combination. The resulting variation can help to ensure that somebody makes it to the next generation.

As in humans, the individual signatures of octopus behavior are not fixed in time. Octopus temperament is malleable from the age of three weeks to six weeks, when temperament (as measured by the three-dimensional scale) varies a lot. Over this period, aggressive animals became shy, and excitable animals became phlegmatic. In humans, personality is most changeable before the age of thirty, after which we tend to settle into a pattern that lasts for many years.

The effects of genetics and environment on personality can be separated by examining animal populations. For example, in dairy goat siblings that were split up into two groups, one reared by humans and the other reared by mother goats, the relative rank order of timidity was the same within each group: the most timid goats in each group tended to be siblings, with a

Did you know? Domesticating the brain

The earliest known domestication of animals dates to more than ten thousand years ago, when dog and human remains first appeared in the same burial sites. It is not known whether domestication initially arose from the gradual selection of more pliant offspring—for instance, by feeding the wolves who were least fearful of a fire—or by selective breeding of captive animals. One twentieth-century experiment by Russian geneticist Dmitry Belyaev suggests that purposeful breeding can lead to especially fast changes in behavior. In his experiment, foxes were selected for docility and only the friendliest pups selected for later breeding. The result of more than thirty generations of such selection was a colony of foxes so friendly that pups actively competed for human attention.

A number of physical attributes often accompany domestication. As Charles Darwin noted long ago, domesticated animals tend to have more floppy ears, wavy or curly hair, and shorter tails than their wild cousins. The recurrent appearance of these traits in different species suggests that breeding for tameness selects for a whole constellation of related traits at the same time. One notable consequence of domestication is relative shrinkage in brain size. In domesticated pigs and chickens, forebrain structures occupy about one-tenth less of the brain than they do in the wild. A mechanism that could account for many of these changes is a tendency for juvenile traits to be retained in adults. In other words, domestic breeding may select for slowed development.

general tendency for the human-reared goats to be less timid. This finding indicates that temperamental traits start from an inborn tendency but can also be influenced by environmental factors such as upbringing.

Many studies of personality and the brain have focused on dopamine and serotonin, two neurotransmitters secreted by cells in the midbrain that are important in regulating the activity of the nervous system. These neurotransmitters are released by nerve terminals throughout the brain and are cleared out by dopamine and serotonin transporters, molecular pumps that suck them back into cells, to be used again or broken down.

How these transmitters are handled can influence the personality of humans and nonhuman animals. For example, in people, studies of identical twins show that about half of the variation in anxiety-related personality traits is inherited. Some of this variation may be due to differences in the action of serotonin. Mice that have been genetically modified to lack a particular type of serotonin receptor show far less anxiety in conflict situations than their normal cousins. In humans, genetic evidence suggests that anxiety-related personality traits may be associated with a shortage of a specific protein that is responsible for the reuptake of serotonin. The influence of the serotonin reuptake protein accounts for somewhat less than a tenth of the inheritable variation in anxiety. In both humans and mice, the relationship between serotonin reuptake and mood can be manipulated. Prozac treats anxiety and depression by inhibiting this serotonin reuptake protein.

Another example of a personality trait that has attracted much research is the tendency to seek out novel experiences. Unsurprisingly, this trait varies inversely with the tendency to avoid harm. Both novelty seeking and harm avoidance are associated with particular types of dopamine receptors, not only in people but also in thoroughbred horses.

Genetic traits such as dopamine and serotonin activity predict personality only to a small degree. The findings are still interesting, though, because they point toward the possibility that we may one day understand how personality traits are determined, both by genes and by environment. Also, even if personality and mood are a black box with a dozen knobs on it, our ability to identify and turn one of them may be behind the effects of drugs such as Prozac.

At the same time, the weakness of these associations does raise the question of how personality can be so visibly inheritable, yet individual genes for personality are so hard to identify. The cumulative picture that emerges from studies of the genetics of personality is that inborn aspects of personality traits are polygenic, meaning that they are constructed from the action of many genes, perhaps hundreds. For this reason, even in the best cases, genetic traits such as particular receptor types have so far only been able to account for a small fraction of the variation seen from individual to individual.

From an evolutionary standpoint, the polygenic nature of personality may be a good thing. Sexual reproduction mixes up the genetic contribution from two parents in unpredictable ways. This allows the dice to be rolled again and again, yielding a range of personalities distributed over the whole spectrum, generation after generation.

The omnipresence of variation in temperament in animals and people leads us to ask whether what we consider abnormal may vary according to the times and local culture. Someone who is considered hopelessly obsessive-compulsive in Papua New Guinea might be a harmless collector of clocks in Switzerland. Even extreme individuals may help our species survive in times of great need. The best warriors of Genghis Khan's army might be locked up as bloodthirsty psychopaths today. So the next time you encounter another diagnosis of attention deficit disorder, just think what a dandy hunter-gatherer that person would have made.

CHAPTER 20

SEX, LOVE, AND PAIR-BONDING

When we talk about mating in nonhuman animals, we're supposed to call it "pair-bonding" rather than love. But if you watched a mated pair of prairie voles together, their behavior would look a lot like love to you. The prairie vole, a small brown burrowing rodent, stays with the same mate for life (which is unusual, since only 3 to 5 percent of mammals are monogamous). Both parents care for the offspring, and prairie voles that lose their mates typically refuse to take another partner.

In contrast, the meadow vole is solitary and has promiscuous breeding habits. By comparing the brains of these two closely related species, scientists have learned a lot about the neural basis of pair-bonding. To measure bonding in the laboratory, scientists allow one vole to wander freely through a container with three rooms connected by tubes. The vole that's being tested is placed in an empty room, which is connected by two passageways to a room containing the vole's mate and another room containing a stranger. The more time the vole spends in the room with its mate, the more bonded it is. Unsurprisingly, the strongest stimulus for formation of a pair-bond is having sex with the partner, but some prairie voles become bonded simply by living together.

Two neuromodulators, oxytocin and arginine vasopressin (AVP), control the formation and expression of pair-bonds in voles. Both these neurotransmitters are important for social recognition in rodents. Oxytocin is released in many mammals during vaginal or cervical stimulation, including childbirth and mating. Oxytocin is important for mother-infant bonding in many species; it seems to be more important for pair-bonding in female voles than in males. On the other hand, AVP is important for a variety of male behaviors—including aggression, scent marking, and courtship—and seems to be the main pair-bonding hormone in male voles. In a pinch, though, either peptide can induce pair-bonding in voles of either sex. Infusion of either neurotransmitter into the brain causes pair-bonding to occur after a short exposure to the partner, even if the pair has not had sex.

The monogamous prairie voles have more receptors for both these neurotransmitters than do the promiscuous meadow voles in certain key brain areas. The two regions that seem to be important for partner preference are in the core of the brain: the nucleus accumbens, which has a high density of oxytocin receptors, and the ventral pallidum, which has a high density of AVP receptors. Locally blocking either set of receptors prevents pair-bonding, as does blocking oxytocin receptors in the prefrontal cortex or blocking AVP receptors in the lateral septum of males. All these areas are considered to be part of the brain's reward system (see chapter 18). Release of the neurotransmitter dopamine within this circuit is critical for the response to natural rewards, like food or sex, and to addictive drugs.

Indeed, love may be the original addiction. Why does the brain have pathways devoted to making people crave white powders that never occur naturally? Perhaps because the brain regions that are important for drug addiction are also the neural circuits responsible for responses to natural rewards, including love. If the ability to become addicted helps animals bond to their mates, perhaps that's why these neural pathways are useful for the survival of the species—and why they persist in spite of the harm that addiction can cause.

Pair-bonds seem to form by conditioned learning, in which the partner's smell (in the case of rodents, at least) becomes associated with the rewarding feelings of sex. This is no different in principle from teaching your dog to sit by associating this behavior with giving him

Did you know? Studying flirtation

Both women and men tend to think of men as the initiators of sexual relationships, but psychologists who study human courtship behavior have disproved that stereotype. Observational study in singles bars (nice work if you can get it) shows that men rarely approach a woman until she has given them a nonverbal signal that it's okay to proceed. Solicitation behaviors—defined as any body movement that caused a male to move closer within fifteen seconds—were mostly unsurprising, and included glancing, primping, smiling, laughing, nodding, requesting aid, and touching the other person. Less attractive women with high levels of solicitation behavior were more likely to be approached by men than more attractive women with low levels of solicitation behavior. In fact, researchers were able to predict whether a woman in a bar would be asked to dance within the next twenty minutes with 90 percent accuracy just by recording how often she did things like glance around the room, smile at a man, or smooth her hair in a ten-minute period.

something to eat—both eating and sex increase the release of dopamine in the nucleus accumbens. Blocking a particular subtype of dopamine receptor prevents the development of mating-induced partner preference, while activating dopamine receptors induces partner preference without mating. After two weeks of bonding with a female, the male prairie vole develops an increased density of another subtype of dopamine receptor that reduces pair-bond formation, presumably to make it harder for him to form a new bond with another female that might interfere with his first pair-bond.

The most convincing evidence that these neural systems account for pair-bond formation is that scientists have succeeded in converting the promiscuous meadow vole to monogamy by experimentally inducing expression of the AVP receptor in its ventral pallidum. This amazing result shows that a complex behavior like pair-bonding can be turned on or off by a single gene in a single brain area, although, of course, other genes in other brain areas are required for the behavior's full expression once the switch is flipped.

Mothers' attachment to their children may involve some of the same neural circuits as bonding with a mate. As we've already mentioned, oxytocin is necessary for mother-infant bonding. When rodents who have never had pups are given oxytocin, these inexperienced females will approach pups and try to care for them, instead of being aggressive toward them as a female nonmother normally would be. Blocking oxytocin receptors during labor and delivery prevents rodent mothers from bonding with their pups. Damaging the ventral tegmental area or the nucleus accumbens, both of which are associated with rewards in female rodents, also impairs their ability to care for their pups.

But enough about prairie voles, cute as they may be. You're probably wondering if this is how people fall in love. We don't know for sure, but there's some evidence that the idea is plausible. Oxytocin levels increase during orgasm in women, and AVP concentrations increase during sexual arousal in men. In addition, functional imaging experiments suggest that romantic love (in both sexes) and male orgasm activate similar reward areas of the brain, the ones that contain receptors for oxytocin and AVP. People who are intensely in love show activity in the ventral tegmental area and the caudate, while people in longer-term relationships (about a year) also show activation of other regions, including the ventral pallidum (the site of prairie vole bonding), when they look at a picture of their lover. These findings suggest that romantic love in humans may involve oxytocin, AVP, and the brain's reward circuitry—all of which are important for pair-bonding in voles.

If you've taken stupid risks when you were in love—and later wondered how you could have trusted that loser—you might be interested to know that oxytocin also seems to increase people's trust during social interactions, even with strangers. Subjects in one experiment were asked to play a game in which an investor could make money by taking the risk of giving some money to a trustee, who would then get the investor's money plus a bonus and could choose how much of it to give back to the investor. If the trustee is trustworthy, both players benefit from the investor's decision; otherwise only the trustee benefits. Investors who were given oxytocin (via a nasal spray) were about twice as likely to give money to the trustee as those who were not given the drug. This effect was only seen when the trustee was a real person, not

Did you know? **Imaging orgasm**

You'd never get this approved by a university in the U.S.: a group of Dutch scientists has been studying human brain activity during orgasm by using positron emission tomographic (PET) scanning. Of course, the brain's reward system is activated during orgasm in both sexes. In addition, women showed reduced activity in an area of the frontal cortex, which might relate to a reduction of inhibition. Men showed reduced activity in the amygdala, suggesting a relaxation of their vigilance during orgasm. Both sexes showed increased activity in the cerebellum, which has recently been implicated in emotional arousal—and sensory surprise.

Myth: **Men learn to be gay**

Research suggests that many homosexual people are born that way—though the evidence is much more clear for gay men than for lesbians. Factors that affect the development of male fetuses also influence adult sexual orientation. Some of these factors are probably genetic, as homosexuality is significantly heritable in human twin studies, while others are environmental influences from the mother during pregnancy. This research doesn't prove that environmental influences after birth are irrelevant, but it does suggest that it is possible to develop a homosexual orientation without learning.

Children with disorders of sexual development provide an opportunity to test this idea because they often have known abnormalities in prenatal hormone exposure. For example, in a syndrome called congenital adrenal hyperplasia, a genetic defect causes female (XX) babies to produce a male steroid hormone that masculinizes their brains and sometimes their genitalia. Even when the hormone defect is corrected with medication after birth, these females are much more likely than normal women to have adult sexual fantasies and experiences involving other women. Women whose mothers took the drug DES, another masculinizing agent, once thought to prevent miscarriages, also are more likely to be attracted to other women, though their genital sex is normal. At the opposite extreme, males with androgen insensitivity have a genetic defect in the receptor for the male hormone testosterone. Because their bodies and brains are not responsive to male hormones, these genetic males (XY) are born with female genitalia and typically raised as girls. Nearly all of them report being attracted to males in adulthood, suggesting that sexual attraction to women requires prenatal brain masculinization by hormones.

If homosexuality is due to early hormones, then we'd expect gay men to look more like women in brain regions that differ between the sexes. In humans, the strongest sex difference occurs in a region with the tongue-twisting name of the third interstitial nucleus of the hypothalamus, which on average is more than twice as large in men as in women. Two studies have reported that this region in gay men is about the same size as in women. As far as we know, no one has studied this region in lesbian brains.

For men without a medical disorder, the strongest known predictor of homosexuality is having an older brother. This effect has been found in more than a dozen studies. Each older brother increases the odds that a later-born male will be gay by a whopping 33 percent. That is, if gay men are 2.5 percent of the male population (which is roughly correct), a boy with one older brother would have a 3.3 percent chance of growing up to be gay, and a boy with two older brothers would have a 4.2 percent chance. According to these statistics, roughly 15 percent of gay men owe their sexual orientation to their older brothers. In contrast, there appears to be no birth order effect on homosexuality in women.

No one is quite sure how having older brothers affects sexual orientation. It isn't because of the mother's age, as this doesn't happen in firstborn male children of older mothers, and it doesn't matter if the older brother is in the house when the younger brother is growing up. The effect probably occurs before birth; homosexual males with older brothers weigh less at birth than heterosexual males with the same number of older brothers. Researchers' best guess at the moment is that the immune system of women who are carrying a male fetus may make antibodies against some factor that males produce, which would then act to suppress that factor in subsequent male babies. One candidate is the Y-linked minor histocompatibility antigen, though the only evidence in its favor so far is from rats; immunizing rat mothers against this protein reduces the likelihood that their male offspring will mate with females and reproduce.

Taken together, all this research suggests that brain development during pregnancy has a significant effect on adult sexual orientation. We can't deny that the expression of people's sexuality also is richly influenced by their life history, but it looks as though the basic plan is laid out early in life.

when a computer randomly decided how much money the investor would get, so oxytocin seems to be involved specifically in social interactions, not in risk taking more generally. These results suggest that you might want to avoid making important financial decisions while under the influence of mind-altering substances, like those released during orgasm.

Life decisions aside, the most dramatic sex differences in the entire brain are found in the parts that control what you do in bed. Here we're not talking about sex differences in cognition, which are subtle and can only be detected by comparing averages for groups (see chapter 25). In contrast, sexual behavior areas of the brain show large enough differences that you can tell whether any particular brain is male or female just by looking at these regions.

These sex differences begin to develop before birth. First a gene on the male-specific Y chromosome directs the production of a factor that induces formation of testicles in male fetuses. The testicles then release testosterone to promote masculinization of the brain and the sex organs, and other hormones to suppress the development of female sex organs. Curiously, female sexual development doesn't require any hormones at this stage, which has led scientists to speculate that female may be the "default" sex.

Aside from a couple of exceptions, hormones act on the brain in two stages. Around the time of birth, hormones organize the brain by controlling the development of regions that will be important for sexual behavior. These behaviors are not expressed, though, until they are activated by male or female hormones after puberty. Both stages must be successful for normal sexual behavior to occur.

Sexual behavior is controlled by an area of the brain called the hypothalamus, which is also important for other basic functions like eating, drinking, and body temperature regulation. In rats, damage to a part of the hypothalamus called the preoptic area prevents male sexual behavior entirely. Several areas of the hypothalamus of rodents show sex differences in their size, with some regions larger in males and others larger in females. For most regions, these size differences are created by hormones during a sensitive period in early life; if hormones are not available when they are needed, these areas will not develop their sex-specific anatomy. However, sex hormones affect the sex-specific anatomy of some regions in adulthood as well, notably a nucleus in the amygdala that is important for male sexual arousal and some regions with AVP receptors that are important for pair-bonding.

As with pair-bonding, we have more detailed information about these pathways in rodents, but there's some reason to believe that the basic system is similar in humans. One reliable sex difference has been found in the human hypothalamus, in an area called the third interstitial nucleus, which is more than twice as large in men as in women. Activation of sexual behavior in adulthood seems to depend on testosterone, the hormone associated with libido in both men and women. Human sexual behavior also depends on a variety of social interactions that are more complex than those of other animals, of course. However, you might be surprised to learn that anthropologists find behavior patterns during flirtation to be very similar across a variety of cultures, suggesting that they too might be strongly influenced by biology rather than cultural experience.

As we've shown, science can explain a lot about love and sex, but certainly not everything. That's fine with us. We're happy to live with a bit of mystery.

PART FIVE

YOUR RATIONAL BRAIN

CHAPTER 21

One Lump or Two: How You Make Decisions

The physicist Richard Feynman was miles ahead of his peers in many ways: he had unmatched intuitions about physical law, he was a lightning-fast calculator—and in his spare time he was a brilliant practical joker. But he had trouble making big decisions, especially when they needed to be settled quickly. He once wrote, "I never can decide anything very important in any length of time at all."

When Feynman joined the Manhattan Project, he encountered a new, critical challenge. Many of the usual rules of academic life—waiting to publish until everything is perfect, proving theorems rigorously—had to fall by the wayside. The crash program to beat the Nazis in the race to build an atomic bomb forced academic physicists to abandon their usual stately pace of progress.

During this period, Feynman was very impressed with a colonel who had to decide whether to allow Feynman to provide a classified briefing to a team at Oak Ridge. The colonel was able to identify the need for a rapid decision—and then make the decision—in five minutes. Once Feynman was cleared to go, he then showed his own particular strength: he told the assembled staff how a nuclear chain reaction works.

Although the conditions of wartime were extreme, decisions are almost always constrained in some way. You rarely have the luxury of all the time or information you want before you make a decision. To take a mundane example, you usually don't know in advance what route will get you to work through morning rush hour most quickly, but you have to pick one or you'll never get there.

Until the last few years, neuroscientists had not studied decision making. Their focus had been on processes more directly related to input (how sensory information is encoded) or output (how actions are encoded). Recently, though, researchers have begun to understand a rudimentary act that comes between input and output: deciding when and where to turn your eyes. This extremely stripped-down version of decision making captures the feature of trading off accuracy against speed.

In such an experiment, a monkey sits in a chair looking at patterns of dots moving around on a screen in front of him. He knows that if he can guess which way most of the dots are moving, the experimenter will give him some juice—orange, his favorite. He peers at the dots, some of which are moving left, some right. It's a confusing mess at first, but he looks a moment longer, then presses a button. Mmmm, juice.

Meanwhile, a researcher sits in the next room, out of sight of the monkey, near a large bank of computers. One video monitor displays the movements of the monkey's eyes, while a loudspeaker clicks in conjunction with electrical signals from neurons in the animal's brain, recorded from an electrode placed in the parietal cortex. The eye movements and neural activity (and juice dispensing, of course) are recorded for analysis later. What is already apparent, though, is that the neuron on the loudspeaker is anticipating the eye movements. The clicks, which represent spikes (see chapter 3), quicken, reach a crescendo just before the animal's eyes move to the right, and then become quieter. Eyes to the left—no change, just a steady low level of activity. A decision to the right—lots of spikes. Over and over again, this neuron's activity presages a decision to look to the right.

The decision-related signals are found in a brain region called LIP (short for lateral intraparietal area). In other brain regions that send their output to LIP, the information about the dots is of a more immediate, sensory nature. LIP seems to integrate the incoming signals to determine which eye movements are more likely to result in juice, though researchers are still arguing over exactly what information it calculates. Delivering small electrical stimuli to LIP can influence decisions, biasing the monkey to look in the wrong direction.

Neural responses in LIP are also affected by manipulations that make the animal more or less motivated. Responses build up more quickly when the animal is paying attention, expecting more juice, or intending to make movements. In each case, neurons in LIP and behavior are affected in the same way. Scientists think that these neurons accumulate evidence of many kinds, and that LIP helps other parts of the brain to decide whether and where to move the eyes.

The neural activity in LIP even reflects the quality of the incoming information. If the dot patterns are less organized, activity speeds up more slowly than with dot patterns that are more clear. A certain level of activity, a "decision threshold," is then reached sooner, allowing a decision to be made more quickly. Thus clearer information leads to more certainty, what engineers call a higher signal-to-noise ratio.

Feynman observed a version of low-noise integration of information when he went to a meeting of a Manhattan Project committee composed of distinguished scientists, four of whom, including Feynman himself, would eventually receive the Nobel Prize. He was amazed to find that debates in this illustrious group could be settled after each member had stated his case exactly once. Anyone who has been through an average corporate meeting can understand why this efficient decision making impressed him.

The simple picture from the monkey experiment, that neurons gather information and figure out when there's enough evidence to stop and choose, might lend insight into the more sophisticated decisions that we humans make. Like Feynman's committee, groups of neurons make decisions by working together to integrate information. Once a threshold amount of evidence is accumulated, the decision is made to move the eyes. However, currently there is no way to observe the interplay among neurons. The nearest anyone has come is to do computer simulations that reproduce what might be happening. In real life, a principal challenge is to find a way to watch the whole group of decision-making neurons at the same time.

Outside the laboratory, decision making is a much more complex business. Human decisions can be about outcomes as large as whether to take a job, or as small as what to have for dinner. In such situations, our brains are called upon to integrate extremely disparate types of information.

Unfortunately, our brains are not naturally equipped to do a good job at integrating complex quantitative facts, probably because they evolved primarily to negotiate social situations and survive natural threats, not to do quantitative puzzles. Classical economic reasoning assumes that individuals are able to evaluate costs and benefits rationally, but the brain's methods of estimation are not good at making such valuations. The payoffs of extremely low-probability events, such as winning the lottery, do not appear to be represented accurately in the brain. If we don't have any intuitive idea of what it means when a probability is below, say, one in one hundred, then the incredible unlikelihood of a lottery payout is not scored rationally. Even though long-

Practical tip: **Maximizers and satisficers**

Both of us have difficulty with decisions. We demand the best outcome, whether we're deciding where to go on vacation or what to have for lunch. That's very hard to achieve. As a result, we're always in danger of spending forever on a decision. For example, when shopping for a plane ticket, we look at dozens of choices, trying to get the lowest fare, the closest airports, the fewest connections . . . Whoops, that one's now sold out. Time to try again. After the decision is made, we waste more time wondering if we were right, which drives our spouses crazy.

Our decision-making style follows a pattern that can be classified as the maximizer model. Maximizers spend a lot of time worrying about differences, no matter how small. In a consumer society with choices everywhere, maximizers suffer from an inability to recognize when an alternative is good enough. Indeed, from an economic perspective, spending the additional time on maximization doesn't make sense since your time itself has some monetary value.

A second category of decision-making style brings more contentment: satisficing, a term that refers to the act of choosing an alternative that is just sufficient to satisfy a goal. Satisficers look until they find something good enough, then stop. Satisficers are decisive, don't look back, and have little regret, even about mistakes. As the saying goes, "The perfect is the enemy of the good." The quintessential example of a satisficer is the Wall Street trader who has to make hundreds of decisions every day and doesn't have the time to second-guess. Psychologist Barry Schwartz has popularized the maximizer-satisficer dichotomy, pointing out that satisficers are, on average, happier than maximizers.

The two of us are slowly getting better at making choices that are perfectly good for the task at hand. Our satisficer spouses are trying to come to terms with our maximizer ways. At least, as satisficers, they are unlikely to question why they married us.

term losses are a virtual certainty, just one anecdotal story of a big winner remains a motivating factor that is weighted out of all proportion to any reasonable expectations. (This is not even to mention that a massive financial reward such as winning the lottery still has only transient effects on happiness, as we explained in chapter 18.)

So people persist in buying lottery tickets, a fact exploited by financially strapped governments everywhere. Even more extreme examples of irrational decision making have been demonstrated. Among the brain rules of thumb explored by Kahneman and Tversky (see chapter 1)

Practical tip: **Can willpower be trained?**

Psychologists have shown that making choices and decisions, making plans to act, and carrying out those plans call upon a resource that can be depleted. In a series of studies done at Case Western Reserve University, people who were asked to do one task requiring an act of will to finish were less persistent in a second task. The two tasks could be as unrelated as eating radishes and working on an impossible-to-solve puzzle. To really drive home the unattractiveness of radishes, they were presented while other subjects received freshly baked chocolate chip cookies. Radish eaters gave up on the puzzle sooner, in eight minutes on average, less than half as long as the subjects who were given cookies. Similarly, subjects asked to perform a boring text-editing task showed less persistence in watching an extremely dull video. Willpower is also reduced after physical exertion or under conditions of stress.

One interesting aspect of the finiteness of willpower is that a variety of tasks call upon the same reserves. Based on this "ego depletion" model, one might expect that exercises that increase willpower in one area might then increase one's capacity to carry out other difficult tasks. Similarly, doing several unrelated tasks in a row that all require active will might be an even more effective means of will "exercise." This is consistent with the sentiment of some psychologists—and self-help books—that willpower is like a muscle. The idea of willpower exercise has culminated in military boot camp, where recruits perform many challenging tasks, and in such spectacles as Watergate-era criminal and maniac G. Gordon Liddy improving his willpower by holding his hand over a candle flame.

Although effortful willpower of any kind interferes with effortful willpower of any other kind immediately thereafter, no one knows why willpower is finite. One possibility is that brain mechanisms for generating active control rely on a resource that can somehow be depleted. Conversely, executive function—the ability to plan and execute a purposeful series of actions—works better if you do it more frequently, which suggests that this resource can grow with practice. One likely place to look is the anterior cingulate cortex, since after damage to this brain region, attention and decision making are impaired.

Broad similarity may exist with other learning systems, which are thought to rely on changes in synaptic connections elsewhere in the brain: willpower-strengthening exercises may cause physical changes in the anterior cingulate and other regions involved in executive function, such as the prefrontal cortex. So practice difficult tasks such as being nice to people you don't like. It might help you stick to that diet.

is that people are notoriously bad at estimation problems. When people are asked to guess the number of beans in the jar, their answer can be swayed by spinning a roulette wheel in front of them while they are thinking about the question, and asking them to consider the outcome of the spin as a possible answer. Despite the obvious irrelevance of this randomly generated number, it can nonetheless nudge the guess upward or downward.

One general principle that has emerged from studies of economic reasoning is that costs and rewards seem to count for less if they are not immediate, and less still if they are in the distant future. This bug in our brain mechanisms has been used to persuade people to save more for their pension funds. In a plan known as Save More Tomorrow, workers are not asked to put away money immediately for retirement, something they are reluctant to do. Instead, they are asked to promise to commit some fraction of their future raises to savings. In this plan, people give up something that they have not yet received. As a result, they do not perceive a loss to their existing lifestyle and are more willing to go along. This is an example of turning a brain bug—the same one that induces you to have bacon now, even if you know it may cause heart disease later—into a feature that works in your favor.

> The essence of ultimate decisions remains impenetrable to the observer—often, indeed to the decider himself.
>
> —John F. Kennedy

CHAPTER 22

INTELLIGENCE (AND THE LACK OF IT)

The idea of intelligence gets people wound up and sometimes defensive, but that's mostly because they focus on the wrong questions. Scientists know a lot about individual differences in intelligence and where they come from, but that information doesn't sell newspapers and magazines. Instead, journalists tend to report on comparisons between groups of people—by gender, by race, by nationality, and so on—and worry that any differences are likely to be used to justify treating people unequally. *That* is the part that gets people wound up.

Intelligence research has a bad reputation, one that was fairly earned by some of the early work in the field. The history of this field is closely tied to attempts to prove that certain groups of people were superior to others and thus deserving of special treatment. In the process, these researchers became the basis of a classic cautionary tale of how biases can influence scientific conclusions.

It is not clear that intelligence has any long-term survival value.

—Stephen Hawking

In *The Mismeasure of Man*, Stephen Jay Gould describes how nineteenth-century attempts to relate brain size to intelligence were compromised by the selection of data to support the conclusions that the researchers knew had to be correct. These guys didn't cheat deliberately; instead, they unconsciously used different standards for data from different groups, which resulted in consistent (and incorrect) findings that their own group had larger brains. Because

Practical tip: How expectations influence test performance

Being reminded of a stereotype just before an exam—even by something as simple as being asked to check a box for male or female—can influence performance substantially. People do worse when they're thinking about a negative stereotype that applies to them, especially when they're told that the task is a difficult one designed to reveal differences between groups. Such effects are seen for stereotypes related to gender, race, age, and socioeconomic status. They can be activated even if test takers are not aware of the reminder, for instance when African-American faces are flashed on a computer screen too quickly to be consciously perceived. Even more curiously, these effects can occur in people who are not members of the stereotyped group: young people walk more slowly after hearing stereotypes about the elderly. This appears to happen because thinking about the stereotype takes up working memory resources (see main text) that would otherwise be used for the test.

The good news is that this problem can be reduced or avoided with a little care. Obviously, teachers shouldn't communicate, directly or indirectly, that certain students are not expected to perform as well as others. Standardized tests should collect demographic information at the end of the answer sheet, not at the beginning. The effect also works in the opposite direction: performance can be improved by exposure to material that contradicts the stereotype, as in girls who hear a lecture on famous female mathematicians before a math test.

Almost everyone fits into more than one group, so perhaps the most practical approach is to bring a more positive stereotype to the task. For example, a mental rotation task shows consistent sex differences, with men performing faster and more accurately than women (see chapter 25). When college students were asked questions that mentioned gender before completing this test, women got only 64 percent as many correct answers as men. In contrast, when they were asked questions that reminded them of their identity as students at a private college, the women got 86 percent as many correct answers as men. The men did better when reminded of their gender, while the women did better when reminded that they were elite students. Thus the gap between men's and women's scores was only a third as large when women were reminded of a positive stereotype that fit them as opposed to a negative stereotype.

Our brains like to make generalizations about groups, as we discussed in chapter 1, so it may be too much to expect stereotypes to disappear entirely. Instead, we recommend taking advantage of your brain's eagerness to take these sorts of shortcuts by choosing the image that suits the way you want to perform. Now that's using your head!

Did you know? **Great brains in small packages**

In 2005, a crow named Betty made the news by constructing a tool. Experimenters challenged Betty and another crow, Adam, to retrieve a bucket from a deep, transparent cylinder. First the birds were given a curved piece of wire, which they used to hook the bucket handle and lift out the reward, a morsel of meat. When given a straight piece of wire, Betty had her insight. She used her beak to bend it into a curve and retrieved her reward. Betty's feat may have been unusually creative for a crow, given that Adam was unable to make the mental leap. But many nonhuman animals engage in complex mental acts.

Among both birds and mammals, some intelligent species come to the front of the class. Parrots, ravens, crows, chimpanzees, and dolphins all have exceptional problem-solving abilities and complex social structures. As we noted in chapter 3, the common feature of mammals and birds with sophisticated cognitive abilities is that a large fraction of their brains are forebrain.

Another impressive feat is the ability to imitate, which requires an animal to observe an action and then translate these observations into motor acts that reproduce it. The nonhuman animals with this skill are great apes (chimpanzees, gorillas, and orangutans), dolphins, corvids (crows, ravens, and jays), and psittaciforms (parrots, budgerigars, and keas). In a typical test, ravens were given a lidded box whose compartments contained pieces of meat. The lids were hinged so that they could be opened by pulling on a flap near the center of the box, but they could also be slid open by pulling sideways on a second flap. Eventually, by trial and error, the birds discovered how to open the box. For a few ravens, the

researchers covered the center flap, forcing the birds to discover the sliding method. If one raven watched another successfully open the box by pulling on the sideways flap, the new raven was much more prone to use the sliding trick.

Finally, large-forebrained animals can create social complexity in the form of larger average group sizes and more complex rules for social hierarchy and interaction. The literal "pecking order" of small-forebrained chickens is an example of a relatively simple social structure. In contrast, large-forebrained animals, like ravens and chimpanzees, live in constantly shifting social groups. We recognize this complexity in our names for animal groups: a parliament of rooks, a congress of baboons.

One group of intelligent animals stands out for sheer weirdness: octopuses. The brain of a common octopus weighs less than a dime and is only half as wide, but the octopus is capable of learning, imitation, puzzle-solving, and deception. For example, octopuses can be trained to distinguish between a red and a white ball. When a trained octopus is placed with one that's new to the task, the second octopus imitates the demonstrator's preference after watching just four times, on average. Octopus keepers often create puzzles for their charges to give them something to do. At the Oregon Coast Aquarium, octopuses had to manipulate a three-part sliding puzzle made of PVC pipe to get at a tube packed with squid—and they did, in less than two minutes.

Invertebrate brains, which are wildly different from those of vertebrates, usually consist of a few clumps of neurons connected to one another by small yarns of nerve. The central brain of an octopus grows by a factor of more than one hundred during the animal's lifetime, a growth rate unmatched in any vertebrate. The human brain is six hundred times the size of an octopus brain, but the octopus also has many neurons in its arms, which may help it to process information.

These observations suggest that the same principles of learning have arisen independently during evolution in invertebrates and vertebrates. Evidently the view that a forebrain is the substrate of intelligence is too parochial. Understanding what octopus, crow, and human brains have in common may help us figure out what it takes to be intelligent.

of the potential for such bias, scientists today often analyze data in a "blinded" fashion, without knowing whether a particular measurement came from the treated or untreated group. In addition, early tests mixed up intelligence with people's knowledge of facts, so that educated test takers did better even if they weren't any smarter than people with less schooling.

These scientific errors mattered because of their effects on public policy. Many early intelligence researchers were drawn to the possibility that humans could be selectively bred like dogs or cattle to create an improved race of man, an idea called eugenics. Of course, how you go about this project depends a lot on your definition of "improved," and it only works at all if the trait that you want to improve depends on genes in some straightforward way. The attempt to breed people for traits like "being respected in society" suffered on both counts. Scientifically, it would have been laughable if it hadn't led to outcomes like forced sterilization of people institutionalized for reasons as diverse as poverty, mental illness, and sexual misbehavior. Many states still have these laws on the books, though they're rarely enforced anymore.

As the study of intelligence has become more rigorous, much work has focused on the factors that affect individual performance. Individual differences in intelligence are much larger than any known differences between groups of people, but one person's intellectual performance can vary over time and across circumstances or tests.

Many subtle situational factors, which are often group-specific, can influence how well someone does on any sort of test. Most people don't appreciate how common or powerful these influences are (see *Practical tip: How expectations influence test performance*). For this reason, although differences in intelligence strongly influence performance across many tasks, these differences are not fixed across the human lifespan. Even more importantly, environmental influences make a strong contribution to the development of intelligence, so group differences that exist in one generation may not carry over to the next. Even if we ignore the ethical problems with the idea, these facts greatly undermine the validity of any attempts to breed people based on the results of intelligence tests.

There are multiple aspects of intelligence, but in this chapter we'll focus on what psychologists call "fluid intelligence," the ability to reason your way through a problem that you've never seen before. This ability is the best general predictor of performance on many different tasks, and it is distinct from the skills and facts (such as vocabulary words) you have already learned. The best measure of fluid intelligence is Raven's Advanced Progressive Matrices, a test that avoids vocabulary discrepancies by using no words at all. Instead, people are shown a set of geometric shapes with common characteristics and asked to choose another shape that fits into the set.

Which parts of your brain are responsible for this ability? The strongest candidate is the prefrontal cortex. Damage to this region leads to difficulty with many forms of abstract reasoning. In normal individuals, prefrontal cortex volume also correlates with fluid intelligence. Finally, the lateral prefrontal cortex is activated by multiple different intelligence tests taken during brain scanning. However, the prefrontal cortex is probably not the only brain region that is

Myth: **Brain folding is a sign of intelligence**

The idea that folds on the brain's surface might be related to brain function dates back at least to the seventeenth century. This idea was further popularized by scientists based solely on the evidence that human brains are more folded than other available brains, such as those of cows and pigs.

The myth was contradicted when several eminent thinkers left their brains to science for measurement after death. Their brains looked very similar to one another, with no physical feature that correlated with intelligence. The distinguished brains were all equally folded and did not look different from less-distinguished brains.

Likewise, in other mammals, brain folding is related not to cognitive sophistication but to absolute brain size. The most folded brains belong to whales and dolphins, the least folded to shrews and rodents. A leading hypothesis of how these folds form is that the connections between nerves pull together the cortical surface, like sloppy stitches bunching up a big sheet. One useful consequence of a folded surface may be to reduce the amount of space taken up by brain wiring: large amounts of axon are not only bulky but also create long distances for signals to travel, making processing times longer. In bigger brains, the cerebral cortex also has more white matter, made up of the axonal wiring that links distant regions to one another. Increased folding and white matter are seen in all large-brained mammals, regardless of their mental sophistication, including humans, elephants, . . . and cows. (The only exception to the rule is the manatee, which has a brain the size of a chimpanzee's but is far smoother. This may be because manatees, otherwise known as sea cows, are unbelievably slow moving and therefore don't need signals to get across the brain quickly, but nobody knows for sure.)

If it's not brain folding, then does brain size determine cognitive sophistication? Not exactly. Brain size depends mainly on body size. Comparing species with one another, brain size increases about three-fourths as quickly as body size. It's not clear why bigger bodies need bigger brains, but one possibility is that the musculature of larger animals is more complex and therefore needs a bigger brain to coordinate movement.

On the other hand, having extra brain mass (relative to body size) does seem to increase cognitive abilities. For example, humans have the largest brains of the animals in our weight class. The extra growth is concentrated in the cerebral cortex: our ratio of cerebral cortex to total brain volume (80 percent) is the highest of any mammal. The runners-up, not surprisingly, are chimpanzees and gorillas.

important for fluid intelligence. Parietal areas of the cortex are also active during many brain-scanning studies of abstract reasoning and intelligence.

Fluid intelligence is closely related to working memory, the ability to hold information in your mind temporarily. Working memory can be as simple as remembering a house number as you're walking from your car to a party, or it can be as complicated as keeping track of the solutions that you've already tried for a logic puzzle while you're trying to think up new potential answers to the problem. People with high fluid intelligence are resistant to distraction, in the sense that they tend not to "lose their place" in what they were doing when they temporarily turn their attention to something else. A brain imaging study found that this improvement was correlated with lateral prefrontal and parietal cortex activity at high-distraction moments in people with high fluid intelligence.

Genes account for at least 40 percent of the individual variability in general intelligence overall, but their influence varies substantially depending on the environment (see chapter 15). Identical twins reared separately after adoption into middle-class households show a 72 percent correlation in intelligence, but this is probably an overestimate of the genetic contribution, since the twins shared an environment before birth (prenatal environment accounts for 20 percent of the correlation) and are often placed in similar homes. Intelligence test results are also strongly influenced by factors like education, nutrition, family environment, and exposure to lead paint and other toxins. Indeed, when the environment is bad, the influence of genes drops as low as 10 percent. Thus, it seems that genes set an upper limit on people's intelligence, but the environment before birth and during childhood determines whether they reach their full genetic potential.

Interactions between genes and environments can be quite complicated, as we've said before. Genetic influences on intelligence become stronger as people get older, perhaps because people seek out environments that suit their genetic predispositions. For example, people with high intelligence tend to be drawn toward professions that require them to exercise their reasoning skills regularly, which may help to keep these skills sharp.

Taken together, this information suggests that proponents of eugenics took exactly the wrong approach to improving human intelligence. As a society, we could increase average intelligence much more effectively by improving the environments of children who don't have the resources to live up to their genetic potential. The controversy over group differences in intelligence distracts attention and resources from a much more productive conversation about how we might do that.

CHAPTER 23

VACATION SNAPSHOTS: MEMORY

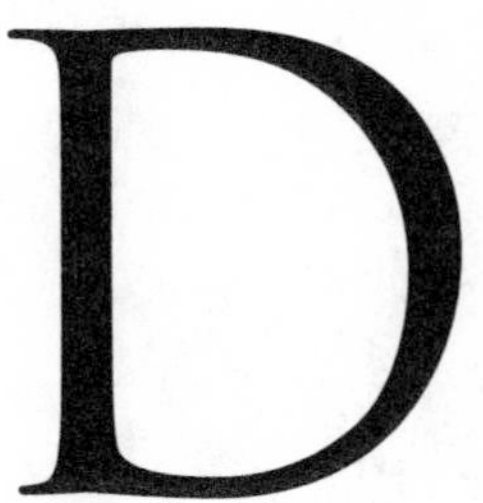

uring most of London's history, which goes back several thousand years, the only way to get around was on foot or by horse-drawn wagon. Because the city was not planned for cars, its roads are a giant jumble. Streets bend and jog and run at odd angles, and they are often narrow, allowing only one-way traffic. Traffic circles and tiny parks are everywhere. Street names change from one block to another. To visitors who are used to streets and avenues organized in orderly grids, it's a mess.

A time-honored way to avoid all this confusion is to hire a cab. Drivers of London's black cabs are legendary for their ability to get to any destination in the city quickly and efficiently. You arrive in Piccadilly Circus, say, and find a taxi. You put all your luggage in the main passenger compartment ("Wow, it's as big as my whole studio apartment in New York!") and give the driver your address, "Grafton Way." After a number of twists and turns—and for most North American tourists, moments of

Did you know? **Forgetting your keys but remembering how to drive**

In the movie *Memento*, Leonard has brain damage that leaves him unable to remember what has happened to him just a few moments before (see chapter 2). This injury makes his life confusing and disjointed. Yet he still remembers how to drive a car perfectly well. How can this be?

Although we commonly think of memory as a single phenomenon, it really has many components. For instance, our brains are able to remember facts (like the capital of Peru) and events (yesterday I had lunch with a friend), and to associate a particular sensation with danger. We also remember how to get to a place in town, how to solve a mechanical puzzle, and how to do a dance step. All these abilities use different brain regions. Together these threads make up the fabric of what we call memory.

Leonard's trouble learning about new facts and events is a defect in what's called declarative memory. This form of memory requires the temporal lobes at the sides of the brain, the hippocampus, and parts of the thalamus, a football-shaped region at the core of the brain.

Other types of memory rely on different brain regions. For instance, the intensity of memory for a terrifying experience like an encounter with an angry bear depends on the amygdala. Learning some types of movement coordination, such as how to make a smooth tennis stroke, requires the cerebellum. A skill such as driving a car uses a number of brain regions but does not require the temporal lobe system, where Leonard's brain damage is located. People with damage to these areas remain capable of learning new skills, like drawing upside down, though they typically have no memory of having practiced the skill before.

seat-gripping fear as you watch traffic rushing at you in the right-hand lane—you are safely at your destination.

The study of the streets of London is a major undertaking that culminates in a daunting examination known as "The Knowledge." Would-be drivers roam London on motor scooters armed with a phone book–sized map, running the maze of streets over and over until they can mentally locate each street and figure out how to get there from any other place. This process culminates in certification exams that require months to take. The average time to be licensed to drive a taxi in London is two years.

Neuroscientists at University College London examined the brains of taxi drivers to see if this intensive study had any effect. The scientists used magnetic resonance imaging to map out the structure of the brains of fifty male drivers and fifty males who did not drive taxis. Only one part of the brain was different in drivers and nondrivers: the hippocampus, a structure that is shaped like a partially unfurled scroll. This difference was small but measurable. The posterior hippocampus of drivers was on average 7 percent larger than in nondrivers, and the anterior hippocampus was 15 percent smaller. Compared with these numbers, the variation within each group is large enough that it would not be possible to tell which group someone belonged to simply by examining the hippocampus. But on average, compared with nondrivers, drivers had a larger back end of the hippocampus and a smaller front end. The more years of experience a driver had, the larger this disproportion tended to be. This difference was not observed in bus drivers, who also drove every day but repeatedly followed the same route. Could it be that acquiring and using The Knowledge makes the hippocampus grow?

What might cause these differences? Active neurons secrete growth factors known as neurotrophins that can cause dendrites and axons to extend their existing branches and even generate new ones. As we mentioned before, neurotrophin secretion is a key event in early development. Similarly, extensive use of neural tissue may lead to growth later in life. New neurons are also born in adults at a low rate, which is higher in the hippocampus than in other brain regions. We don't know for certain how the expansion in size and number of neurons would affect function, but a plausible guess is that it would expand as well.

This leads us to one of the core questions in neuroscience: what is it that changes in the brain when we learn something? The difficulty is that few of these changes are likely to be visible when we look at gross structure. Instead, new information is likely to be stored as changes in the strength of connections between neurons, and as changes in which connections are made. These changes don't necessarily alter the size of a brain structure any more than the size of a piece of paper changes when you write on it. So measurement of the sizes of brain structures is a fairly crude and indirect way to assess their capabilities.

The original reason that these researchers decided to look at the hippocampus was that it is known to be involved in spatial navigation in humans and in other animals. As rats run around in a maze, neurons of the hippocampus fire only when the rat is in a particular location. Because the rat hippocampus contains millions of neurons, each place in the maze is then associated with hundreds or thousands of neurons that fire when the rat is there, but not before or after. Taken together, all the neurons of the hippocampus, firing and not firing, hold a map composed of place cells, in which subsets of firing neurons signify where the rat is.

The same phenomenon has been found in humans during a video game that is very similar to what London taxi drivers do every day. Recording from individual neurons in humans is normally not advisable because it requires opening the skull, but it has been done in people with severe epilepsy. In these patients, electrodes are often implanted to identify places in the brain where seizures begin, so that those parts can be removed without damaging neighboring regions important for normal function. Researchers took this as an opportunity to spy on the activity of neurons as patients played a taxi-driving video game. The game involved driving to various destinations in a simulated town, like a very boring version of Grand Theft Auto, without the gangs, crime, or sex.

Very much like in rats, the hippocampi of human virtual cabbies had place cells. For instance, some cells fired when the player was in front of the drugstore, but did not fire when he was at the grocery store. The specific response of cells to various imaginary locations began after subjects had played the game just a few times. How does this happen so quickly? One possibility is that something like a blank map is already in place in your head, waiting to be linked up to experiences of actual places. This may be the first step in learning to navigate a new locale—like a taxi driver in training going around on a scooter with a map.

In addition to being involved in forming memories of places, the hippocampus also is important for declarative memory (the recall of facts and events). For example, if you remember the taxi trip through London earlier in this chapter (and we hope you do), you are using declarative memory. Canadian psychologist Brenda Milner was the first to appreciate the importance of the hippocampus and structures near it for this form of memory. In the 1950s, Milner examined a patient, HM, who had undergone radical surgery to treat severe epileptic seizures. Like the patients who played the taxi-driving game, HM's seizures began in the hippocampus or nearby in the temporal lobes of the cerebral cortex. However, at that time it was not standard practice to record activity before surgery. Doctors only knew that seizures often began in the temporal lobes and the hippocampus. So they surgically removed these structures in their entirety.

After the surgery, HM's seizures were indeed less frequent. He was also able to have conversations, solve logic puzzles, and carry out the activities of daily living. But he had an odd deficit as well. He suffered a profound loss in his ability to remember an event, even a few minutes after it happened. Milner tested him many times over the following months. He did well on the same tasks and even improved with repetition. Yet he could not form new memories of events or people. For instance, each day he greeted Milner as if meeting her for the first time.

Milner and other neuroscientists eventually reached the conclusion that temporal structures are essential for forming declarative memory. The problems experienced by HM have now been seen in many patients after strokes have damaged their temporal brain structures, including the hippocampus.

Since both place memory and episodic recall require the hippocampus, scientists speculate that these two forms of memory may share some common principle. One idea is that they both

Myth: Recovered memory

Memories are not played back like a tape or a file recalled from a computer's hard drive. Instead, they seem to be stored in shorthand, broken into chunks in which the uninteresting bits are discarded, leaving only the details that your brain considers important. As we discussed in chapter 1, your brain also invents details to create a more coherent story. This has occasionally caused memorable tragedies.

In a wave of scandalous cases in the 1980s and 1990s, social workers and therapists identified "repressed memories" of childhood abuse. The stories were uncovered after interviewers repeatedly asked leading questions and then rewarded the most interesting answers with attention. In Manhattan Beach, California, a lawsuit claimed that hundreds of children had been sexually abused at the McMartin Preschool, some in nonexistent networks of underground tunnels. These unbelievable tales led to lengthy court cases and the wrongful imprisonment for five years of Ray Buckey, a counselor at the school.

Filling-in of memories is a well-documented phenomenon. In one study, researchers asked people where they were when they learned that the space shuttle *Challenger* had exploded. People gave different answers several years later than they did immediately after the explosion, providing more evidence that people sometimes invent plausible explanations when they don't recall what happened.

Researchers have stimulated false memories in the laboratory as well. For instance, if you are shown a list of words with a similar connotation—*ice cream*, *honey*, *lollipop*, *sweet*, *candy*, *chocolate*—and later asked if the word *sugar* was on the list, there is a good chance that you will say yes with confidence. This is an example of filling in, in which a reasonable inference is made that an event might have happened, even though it did not.

The fragility of memory plays into another common myth, which dates to the teachings of Sigmund Freud. He speculated, without hard evidence, that traumatic events could be repressed and thereby made unavailable to the conscious mind. The concept has become so entrenched that it is believed today, even by many mental health workers. However, almost no scientific evidence exists for repression. The weakness of the evidence is detailed in psychologist Daniel Schacter's *Searching for Memory*. Severely traumatic experiences are forgotten only if the trauma leads to unconsciousness or brain damage, or if the experience happens to a person too young to be able to form long-term memories, a process that begins around the age of three or four. Most memory researchers agree that recovery of a lost traumatic memory is very rare.

Practical tip: **Can't get it out of my head**

Anne Waldman is stuck. She and her son are hard at work on a collection of songs based on her poetry, entitled *The Eye of the Falcon*. As she polishes the songs, she finds that she can't get a certain phrase out of her head. It's driving her crazy. Why is this little phrase so persistent?

Think of the phrase bothering her as an example of a sequence. Sequence recall has a special and useful place in our memories. We are constantly called upon to remember sequences, from the movements involved in signing your name or making coffee in the morning to the names of the exits that come before the turnoff you take to drive home every day. The ability to recall these sequences makes many aspects of everyday life possible.

As you think about a snippet of song or speech, your brain may repeat a sequence that strengthens the connections associated with that phrase. This in turn increases the likelihood that you will recall that phrase, which then leads to more reinforcement. This cycle of repeated recall may be necessary for the normal strengthening and cementing of memories.

In Anne's case, though, the repetition helped form a positive feedback loop and a vicious cycle. At first she recalled the phrase on purpose, but after a time it arose unbidden. In her case, the bothersome phrase is one she was actively working on, and one with substantial emotional impact. Emotions can highlight the effect of experience and make events more likely to be consolidated in memory.

How can one break this unending cycle of recall and reinforcement? One way is to introduce other sequences that interfere with the reinforcement of the memory. Thinking of another song may allow a competing memory to crowd out the first one. Anne attempted to overwrite her repetitive memory by listening to a Poulenc opera with Jean Cocteau. That's the best therapy we can suggest: find another infectious song—and hope that the cure doesn't become more annoying than the original problem.

rely on placing events relative to one another in context. In spatial memory, the relationship is physical, in space; in episodic memory, the relationships are more general, in time or even by logical connection. What physical property of the hippocampus allows it to make these logical connections?

About a hundred years ago, the psychologist William James suggested that our experiences trigger sequences of activity in the brain. Under the right conditions, these sequences can then lead to changes that increase the likelihood that they will occur again, even after the experience is past. If the activity sequence is repeated, eventually the change becomes strong enough that the entire sequence can be triggered by some cue that evokes the beginning of the sequence.

In 1949 the Canadian neuropsychologist Donald Hebb suggested how James's change might take place. He proposed that the essential components of learning were the firing of neurons in a precise order, and the connections between them, synapses, that set up the order. In his formulation, the strengthening and weakening of synaptic connections between neurons could be the underlying means by which a sequence of neuronal firing is reinforced. More than twenty years after Hebb made this suggestion, Terje Lømo and Timothy Bliss proved him right. They found that synapses could indeed change their strength in a lasting way after being activated (as we discussed in chapter 13). This phenomenon, called long-term potentiation, has since been found in a variety of animals, including primates, rats, rabbits, slugs, insects, birds, and even octopuses. These changes last for minutes to hours. On longer timescales, connections may rearrange themselves and new ones may grow, perhaps even leading to structural changes like those seen in the London taxi drivers' brains.

How do these ideas apply to the hippocampus? Many neurons in the hippocampus excite other neurons nearby, so one neuron can excite another, which excites the next, and so on—perhaps in long sequences, all within the hippocampus. This sounds very much like Hebb's vision of sequences of activity as a means of reliving an experience. Perhaps the hippocampus's internal loops of excitation allow these sequences to be generated.

These loops of excitation might also play a part in why the hippocampus and temporal lobe are so prone to epilepsy. If these structures have a tendency to form positive feedback loops, then they might be likely to initiate epileptic seizures, which are periods of runaway brain activity. Indeed, the cerebral cortex is also full of internal connections—and the cortex is another major site for seizures to begin.

CHAPTER 24

RATIONALITY WITHOUT REASON: AUTISM

If you've spent much time reading newspapers and magazines over the past few years, you may have formed the impression that autism is caused by environmental toxins of some sort, perhaps by vaccination. According to one recent analysis, this idea receives seven times as much attention in the popular press as it does in the scientific literature on which press accounts are ostensibly based. Although it makes a good story, the environmental hypothesis does have one major drawback: it's most likely wrong—or at least incomplete.

"Autism" is a catchall term for a highly variable set of behavioral disorders that begin in early childhood. It is defined by three features: lack of social reciprocity, disrupted verbal and nonverbal communication, and inflexible and repetitive behaviors. Autism affects six out of a thousand people today and is four times as common in males as in females. People who have normal language but exhibit the other two features are diagnosed with a related disorder, Asperger's syndrome.

The social behavior problems caused by autism are very distinctive. One way of describing these problems is in terms of what researchers call "theory of mind." This phrase refers to the human ability to imagine what other people know and what they are thinking or feeling, an ability that develops in most children around the age of three or four. People with autism have extreme difficulty imagining anyone else's point of view, and consequently have trouble recognizing when others are lying, being sarcastic, mocking them, or taking advantage of them. They have particular trouble with responding appropriately to faces, including recognizing or remembering them, as well as detecting facial signals of emotion. Most people pay the most attention to the eyes when looking at a face, but autistic people tend to look at the mouth or elsewhere in the room.

Sam grew up with an autistic younger sister. As a small child, Karen was late to start talking. As a toddler, she was prone to hitting other children and shouting at inappropriate times. Talking with her was an exercise in frustration. She responded to questions such as "How are you?" by repeating the question, and when prompted to give an appropriate answer ("Karen, say you are fine"), she replied, "You are fine"—creating endless frustration for both parties. Easily overstimulated, she spent a lot of time sitting in a corner tapping one finger repeatedly against a finger on the other hand. This form of self-entertainment seemed to soothe her but was not exactly conducive to group play. As a boy, Sam didn't like to have friends over for fear of being interrupted by bizarre yelling or something worse. He found friends' houses or the library to be more peaceful than home.

Karen's problems were apparent enough that she was diagnosed as autistic by the age of five, which was an early diagnosis in the 1970s, before autism became a well-known disorder. At that time autism was even less understood by the public than it is now. Her parents spent decades thinking something had happened to her in early childhood to cause her autism. For example, she was born prematurely, and they thought her problems might have been caused by rough handling as a newborn, when the plates of her skull had not fully closed.

A feeling of responsibility or self-blame is common among parents of autistic children; this feeling has its roots in the assumption that the disorder must have an environmental cause. For many years, psychiatrists attributed autism to the emotional coldness of "refrigerator mothers"—a complete misunderstanding, but one that fit well with parents' feelings of responsibility. In general, diseases that are not well understood often acquire a reputation of being caused by the environment. Another example is ulcers, which were long thought to be caused by stress but are in fact caused by bacteria.

We don't know exactly what causes autism, but we do know that it is a disorder of brain development with a very strong genetic component. If one of a pair of identical twins has the disorder, the other twin has a better than 50 percent chance of being autistic, even though twins in general are not at higher risk for autism than single-born children. Even nonidentical siblings of autistic children have twenty-five to sixty-seven times more risk of autism than the general population. And relatives of autistic people have a higher chance of having some autistic symptoms even if they are not fully autistic.

However, despite the strong contribution from genetics, there is not a single "autism gene." There are a few rare syndromes in which autistic symptoms can result from a mutation in just one gene. But in most cases autism requires some combination of genes to be present. We know this because pairs of fraternal twins, who share half their genes with one another, have at most a 10 percent chance of sharing an autism diagnosis. This tells us two things: First, because the

Did you know? **Monkey see, monkey do: Mirror neurons**

Social skills depend on empathy, the awareness of what others are feeling. Empathy is not present at birth but must be developed in childhood. Studies in psychology suggest that imitation is one way that children learn to read body language and facial expression in others. Young children tend to imitate others as if looking in a mirror, moving their left hand when someone else moves his right hand, and they also tend to imitate the goals of an action rather than the action itself.

Neuroscientists have found brain circuits that are specialized for imitation and may also be important for empathy. What researchers call "mirror neurons" are found in the inferior frontal gyrus and premotor and parietal cortex in monkeys. They are active when the animal performs a goal-directed action, such as grasping food, or when he watches another animal perform the same action. Some mirror neurons are active only when the animal sees someone else make the exact same movement, but others are active when someone else achieves the same goal in a different way. Some mirror neurons are even activated by a sensory stimulus that suggests an action that cannot be seen, like the sound of a piece of food being unwrapped or the sight of a hand disappearing behind a barrier where the monkey knows there is food. Mirror neurons also seem to distinguish the intention behind a given action, so that a particular neuron might fire when food is grasped by someone intending to eat it but not when it is grasped by someone intending to put it away in storage.

These two areas are also active during imitation in human brain imaging studies. Magnetic stimulation that disrupts the function of the inferior frontal gyrus interferes with imitation in humans. A major input to the parietal mirror neuron region is an area called the superior temporal sulcus, which is important for attributing mental states to other people. In normal ten-year-old children, the mirror neuron areas are more active in individuals with higher scores on a test of empathy, suggesting that empathy may be learned by imagining yourself in other people's shoes.

The social deficits seen in autism may involve a dysfunction in the mirror neuron system. Autistic children show less activity in these brain areas than normal children when asked to observe or imitate facial expressions. In addition, the decrease in activity correlates with the severity of the autistic symptoms. Of course, these findings do not prove that deficits in the mirror neuron system cause autism, and there are many other brain regions that do not respond normally in this condition, including the brain area that is specialized for face recognition. Another possible site for problems in autistic people is the insula, which is active in processing both one's own emotional state and that of others (see chapter 16). These promising ideas will attract much more research over the next few years, which should give scientists more clues about the causes of autism.

environment is likely to make a similar contribution for both fraternal and identical twin pairs, the effect of environmental causes must, on average, be weak. Second, the chance that two fraternal twins are both autistic is far lower than the odds for identical twins. This is a typical pattern of inheritance for a disorder that depends on multiple genes. To take a simple example, if someone's autism is caused by inheriting two different genes containing mutations (let's say gene A from the mother and gene B from the father), then there is only one chance in four that the sibling of the autistic person will have exactly the same copies of both gene A and gene B. For more genes, the chance is even lower. This sort of analysis has led scientists to conclude that most autism is caused by mutations in two to twenty genes.

Even if autism turns out to be entirely caused by genetic mutations, though, that still leaves open the possibility that it can be influenced by the environment. A good example of an interaction between genes and the environment is another disorder, phenylketonuria, which results from a genetic mutation that disrupts the function of the enzyme that converts the amino acid phenylalanine to another compound. When phenylalanine builds up in the body, it damages neurons, causing mental retardation and permanent behavioral deficits. This damage can be prevented by an environmental manipulation—removing all phenylalanine from the diet.

One argument that seems at first glance to favor an environmental cause for autism is the increase in diagnosed cases over the past four decades. The numbers seem impressive: there

Myth: Vaccines cause autism

The proposed link between vaccines and autism has received a lot of attention over the past few years. Robert Kennedy Jr. wrote a book about it. Indiana Republican Dan Burton, who has an autistic grandchild, has held multiple congressional hearings on the topic. Scientists have spent hundreds of hours and reviewed thousands of patient records to investigate this connection, but have found no trace of a causal relationship—still, the speculations continue.

All this excitement began with a 1998 study from a British gastroenterologist. The paper reported on twelve patients, who were selected based on gastrointestinal symptoms. Nine met the criteria for autism diagnosis. The parents of eight of the children reported that the symptoms had begun around the time that the children were vaccinated against measles, mumps, and rubella (known as the MMR vaccine). The paper noted that the behavioral and intestinal symptoms may have occurred together by chance, "reflecting a selection bias in a self-referred group."

The paper's interpretation was later retracted by ten of the gastroenterologist's twelve coauthors, who stated, "We wish to make it clear that in this paper no causal link was established between MMR vaccine and autism as the data were insufficient." Indeed, the study did not even have a control group, which was essential considering that the outcome measure, intestinal inflammation, was so vague and common. Others were not able to reproduce the gastroenterologist's findings. It also came to light that, before the paper's publication, the lead author had been consulting for a group of lawyers who were intending to file suit against vaccine manufacturers.

Parents may associate vaccination with the onset of autism by coincidence because both events occur around the same time. Vaccines are given between twelve and fifteen months, and symptoms of autism typically begin to appear between twelve and twenty-four months. In one study, starting in 1979, all the cases of autism or autism spectrum disorders in a London district were identified. Autistic children were no more likely to have been vaccinated than typical children. The diagnosis of autism was no more likely to occur immediately after vaccination than at any other time. A study in Sweden also found that introduction of the MMR vaccine did not correlate with an increase in autism diagnosis. Indeed, in several independent reviews by the U.S. Institute of Medicine, the U.K. Medical Research Council, and the Cochrane Library (an international consortium of scientists formed to evaluate the medical literature), no credible link between vaccines and autism has been found. The Cochrane group notes that most studies on the subject are flawed by unreliable outcome evaluation and other sources of investigator bias.

The hypothesis favored by Kennedy is that autism is caused by ethyl mercury in thimerosal, a preservative that was used in some vaccines (though not the MMR vaccine) until 2001 in the U.S. The main evidence for this idea is that autism diagnosis has been increasing over the last few decades, though it is unclear whether this reflects a real increase in the number of affected people, as discussed below. Even if we accept that there is an autism epidemic, though, it does not correlate with the presence of thimerosal in vaccines. In the London study, no jump in autism diagnosis occurred when vaccines containing thimerosal were introduced in 1988. Thimerosal was present in U.S. vaccines between 1991 and 2001, but the increase in autism diagnosis began earlier and has not declined since the preservative was removed. Canada and Denmark removed thimerosal from their vaccines in 1995 and have since had no decrease in the rate of autism diagnosis. Sadly, the continued debate about this false trail diverts needed resources from productive lines of research into the true causes of autism.

has been a fifteenfold increase in the reported prevalence of autism since the first studies in the 1960s. On closer inspection, though, several important factors have changed between early and contemporary studies. First, the diagnostic criteria are different now, and even a small change in the criteria leads to very large changes in the measured prevalence. Many kids diagnosed with autism today would not have qualified when the first criteria were formalized in 1980. Many people who are now diagnosed with autism would previously have been institutionalized, while others might have been neglected, living without any meaningful integration in their communities. Second, parents and doctors know more about autism and are more likely to consider the possibility when evaluating a child with developmental problems. Third, better treatment options are available, which increases parents' motivation to identify autism in their children. Many parents are interested in obtaining behavioral therapy, which, although not curative, may lead to some improvement in symptoms. Of course, by the same arguments, no one can say for sure that the prevalence of autism has not increased. In fact, some scientists believe that autism is underdiagnosed even now. What we can say is that the data from past decades doesn't provide clear evidence for an increase in the prevalence of autism.

Whatever the relative importance of genes or the environment in causing autism, both act by affecting brain development. The brains of most autistic people do not appear dramatically different from normal brains, though some autistics have unusually large brains and, for unknown reasons, unusually small cerebellums. These differences in brain size are not present at birth but instead develop over the first two years of life, suggesting a problem with the "pruning" of brain connections that normally occurs during this time period, as we discussed in chapter 10. Most autistic people have a variety of subtle but widespread problems in the cortex and other areas, including changes in neuron density or number, and the disordering of the normal arrangement of neurons into functional groups.

Only a few specific genes have been consistently linked to autism. If multiple mutations are required to cause the disorder, then geneticists may never be able to identify all the complicated interactions that are involved. Even partial answers can be useful, however, in suggesting brain mechanisms of the disorder. For example, autism is linked to mutations in two families of related genes, called neurexins and neuroligins. These genes encode proteins that control the positioning of neurotransmitter receptors during the formation of both excitatory and inhibitory synapses in early development.

This is interesting because about 30 percent of autistic people also have epilepsy, compared with only 1 percent of the general population. Epilepsy is a disease of brain excitability that occurs when the balance between excitation and inhibition is disrupted, leading to uncontrolled excitation that causes seizures in the body. It is easy to imagine how damage to the neurexin or neuroligin genes could lead to defects in this synaptic balance that cause seizures. It is not much more difficult to picture such changes causing more subtle functional defects in brain regions that control language or social behavior, though no one is certain exactly how this happens.

Some scientists suspect that all these differences between autistic and normal brains result from a primary defect in connections between brain areas. In particular, many autism symptoms could be explained by damage to connections that allow the frontal cortex and other so-called association areas (which coordinate the use of many different types of information) to influence brain regions that are important for routine behavior and sensation. Without these connections, the brain would be unable to regulate incoming sensations, which could cause the hypersensitivity to environmental stimuli seen in many autistic people. The association areas are also important for facilitating flexible responses to circumstance, including suppressing habitual behaviors when appropriate in a particular context, which could account for rigid and repetitive behavior. Finally, many of these association areas are directly involved in social behavior (see chapter 16).

One question is why the genetic factors that underlie autism would persist in the population.

It's possible that individually, the genes confer some benefit. For example, autistic people tend to be very good with details, perhaps because of a lack of higher control from the frontal cortex. A small number of people in the population with an exceptional ability to focus on tasks could be a good thing for society. In the words of the famous autistic Temple Grandin, "What would happen if the autism gene was eliminated from the gene pool? You would have a bunch of people standing around chatting and socializing and not getting anything done."

CHAPTER 25

A Brief Detour to Mars and Venus: Cognitive Gender Differences

Men and women are exactly the same.

Just kidding. If we had to toe that ideological line, this would be a very short chapter. Now, it is true that many sex differences are exaggerated—and some are just plain invented. The world is full of nurturing men and aggressive women, and the sexes are equally smart overall. But as anyone who's raised kids probably knows already, boys and girls are born with some different equipment between their ears.

Of course there are major differences in the brain regions that determine which sex you'd rather see in tight pants (see chapter 20). But get your mind out of the gutter for a moment, and let's consider why men and women might think differently when they're not in bed. We know that hormones influence how the brain works and that sex hormones like testosterone and estrogen are present in different amounts in males and females. These hormones have an especially strong influence before and soon after birth, when babies' brains are being formed, but they also have direct effects on adult brains. Men's and women's brains are shaped differently too, probably as a result of these hormones—though again, most of the differences are very subtle. Women's brains have slightly more surface area and more connections between areas, while men's brains have slightly more volume, even when we allow for their larger bodies.

Given these differences, it's hardly surprising that men and women might tend to behave differently. But human behavior is determined not only by biology, but also by experience and training—what we commonly call culture. Most kids want to behave in ways that please their favorite adults. If girls are punished for getting their clothes dirty, while boys sense that their

Myth: Women are moodier than men

We can't deny that women are moody. What most people don't realize is that men are moody too. In fact, their moods vary as much from hour to hour as women's moods. How do we know this? When psychologists give beepers to men and women and ask them to write down their mood whenever it goes off, men and women report similar variations. Curiously, both men and women tend to remember women's mood swings better, so if people are asked later to remember how moody they or their partners were in the previous week, more mood swings are reported for women than men.

It is true that mood disorders, including depression and anxiety, are about twice as common in women as in men. Some of that disparity may be because women are more willing to go to the doctor when they feel bad, but even when we account for that cultural difference, women are still at greater risk. No one is really sure why, though some people have guessed that women's life experiences may expose them to more stress, which is linked to depression and anxiety (see chapter 17). Men and women are equally prone to manic-depressive disorder, which is strongly linked to genes.

parents are secretly happy about such a show of masculinity, then we can't conclude that girls are naturally inclined to be fussy about their appearance. Many teenagers believe that men find smart women less attractive (though thankfully most of us come to know better), and the effectiveness of all-girls' schools in promoting academic achievement suggests the possibility that girls may adjust their behavior—and their apparent abilities—to accommodate this stereotype.

Prior beliefs can also influence how people's performance is evaluated by others. Beginning in the 1970s, debate raged in the classical music community over whether women could play as well as men, since the top orchestras were made up overwhelmingly of men. Then feminists convinced U.S. orchestra directors to start having musicians audition behind a screen so the judges could hear the music without seeing the player. Surprise! Twenty years later, half the players in the top five orchestras in the U.S. are women. In Europe, though, blind auditions are rare, orchestras are largely male, and many musicians still believe that women can't play as well as men.

So how do we distinguish between biological and cultural influences on behavior? We can't separate the two absolutely, since the environment shapes the way our brains work, but we can make educated guesses. For instance, behaviors that differ between males and females in other species are more likely to reflect biological differences. (Rats, for instance, don't have much culture.) Behaviors that are reliably masculine across various cultures are also more likely to have a biological basis (though the biology in question could be men's greater muscle strength, not necessarily their brains). With that in mind, let's look at some of the more convincing biological sex differences that are documented in people.

The most reliable difference is in spatial reasoning. Not that men don't like to ask for directions—that's probably cultural—but that males typically think about the physical arrangement of the world differently then females. Even female rats depend more on local landmarks to find their way around, while males work from a mental map of space. For example, consider a maze in which the path to the reward can be memorized by paying attention either to local features on the walls of the maze or to distant features on the walls of the room. Rotating the maze so that it faces a different wall of the room (which changes the distant cues) doesn't affect the performance of female rats much, but it causes the males to make mistakes. Changing local features affects performance more strongly in female rats than in males. Similarly, if you hear someone say, "Go past the stone church on the left, and then turn right a few blocks later at the tan house with the big pine tree," you're probably listening to a woman. If you hear "Go south for 1.6 miles, then go east for another half mile," odds are it's a man talking.

These differences extend to more abstract forms of spatial reasoning. For instance, starting with an unfamiliar object photographed from one angle, men are faster and more accurate than women

Did you know? **Males are more variable than females**

People tend to focus on the fact that more males than females score extremely well on math tests, but it's also true that more males score very poorly. Indeed, male scores are more variable than female scores across many tests of mental abilities. This is another way of saying that more males than females have abilities that are far from the average in both directions. Like most sex differences, this one is small, and only becomes important for individual people who are very far from average.

One possible evolutionary reason for this difference is that females are more important to the production of children. If some males in the population go out and get themselves killed, or fail to reproduce for any reason, the total number of children may be unaffected because the remaining men can make up for the losses. On the other hand, if there are fewer women, there are likely to be fewer children. That means that genetic changes that lead to higher variability between individual men are more likely to survive in the population because they're more likely to be passed on to the next generation.

at deciding whether a second picture is the same object seen from another angle. We know this difference is probably due to hormones for one simple reason: if you give testosterone to women, they suddenly get a lot better at the task. (In the long term, they also grow chest and facial hair, so this isn't a great solution for most women.)

The sex difference in mental rotation tasks is large, with the average man performing better than about 80 percent of the women. For comparison, though, even this cognitive difference between the sexes (one of the largest known) is smaller than their difference in height: a man of average height in the U.S. is taller than 92 percent of the female population.

Men aren't better at all spatial reasoning tasks, though. It's not a coincidence that the woman of the house knows where the mustard has ended up in the back of the fridge, but a reliable sex difference. (You can try this with your friends: arrange a set of ten or twenty objects on a tray, let everyone look at them for one minute, then rearrange the objects and ask everyone to write down which objects are in a new location.) Women are better than men at remembering the spatial location of objects, and their advantage at this task is as strong as the mental rotation advantage for men.

What about intellectual abilities? In 2005, Larry Summers, the president of Harvard, got himself into a lot of trouble by saying in public that men are better at math than women. To be fair, what he actually said was that more men than women have very, very high scores on standardized math tests. It would be nearly impossible to look at a person's math scores and decide whether the test taker was male or female because there's a huge overlap in abilities for most of the population. But among the very highest (and lowest) scorers on math tests, men outnumber women dramatically. This imbalance between the sexes might be a biological difference related to the male advantage in abstract spatial reasoning, but it might also be the result of our culture telling women that they aren't good at math. For example, it's possible to lower women's math test scores just by asking them to write down their gender on the first page of the exam (see chapter 22) or to raise them by asking women to think about high-achieving women before taking the test. (Please do try this at home!) In addition, test scores don't predict academic performance very well; in fact, males tend to do worse in college math classes than their test scores would predict, while females tend to do better. So the jury's still out on whether the sex difference in high-level math scores involves differences in men's and women's brains or in their cultures.

As long as we're being politically incorrect, the other place we find a lot more men than women is in prison. Men are much more likely to get into trouble for violent behavior. That could mean that men's brains are biologically more inclined toward aggression, or it could just mean that men are big and strong, so they're more likely to use violence because it's effective for

Quiz: How to think like a man

Which one of the three comparison shapes on the right is a rotated version of the standard object on the left? Answer as fast as you can, using a watch with a second hand to time yourself. (The answers are at the end, but don't cheat!)

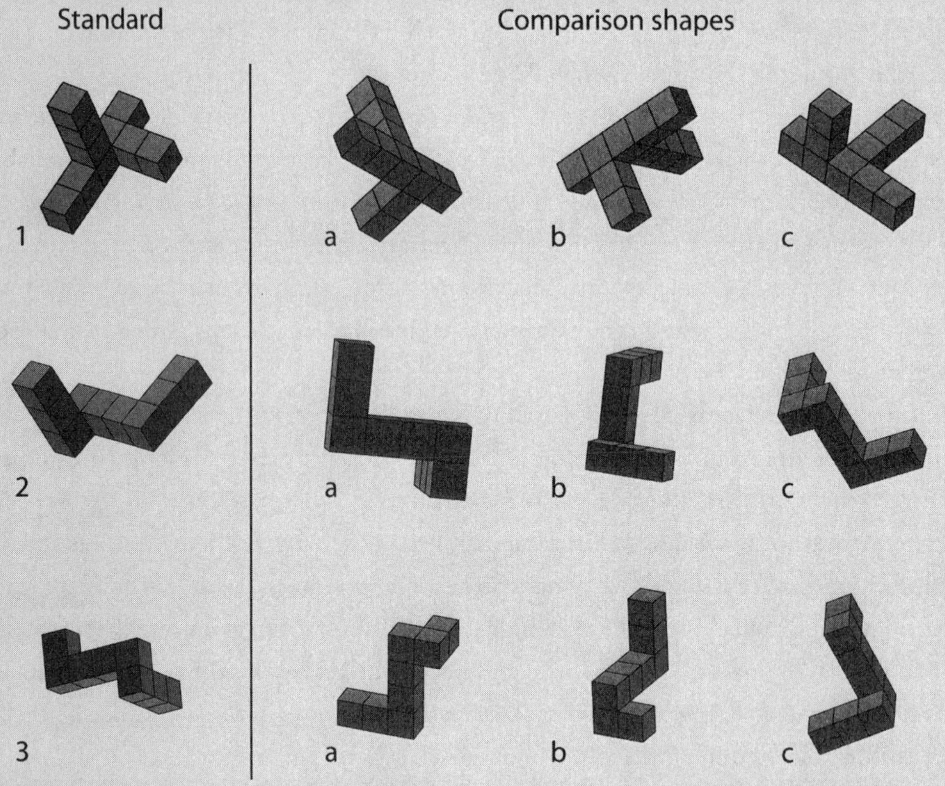

Standard — Comparison shapes

1 — a, b, c

2 — a, b, c

3 — a, b, c

This test demonstrates one of the largest known differences between men's and women's brains. We know a neuroscience professor who grew up as a woman and eventually realized that she had always felt like a man and wanted to change her gender. Being a scientist, she signed up for a study on sex differences in cognition that focused on mental rotation of objects, like the test above. Running the study during the sex change gave the researchers the unusual advantage that they could use the same person in the female (before) and male (after) groups! Before the hormone treatment, our friend found the test quite difficult, and felt that she had to slowly rotate each shape in her mind to see if it matched the standard. After the testosterone injections started, the test got easier and easier. By the end of the study, as a man, the correct answer seemed immediately obvious. This is the clearest description we've ever heard of what this sex difference feels like from the inside.

Answers: 1) b, 2) a, 3) c

them. Aggression is more socially acceptable in boys, but that's not the whole story, since many modern parents have found to their dismay that boys have a stronger tendency toward violent play than girls do, even when the parents are determined to treat their sons and daughters in the same way. Young male monkeys also engage in more rough play than their female counterparts—and even prefer toy trucks over dolls. Although aggression levels vary enormously between different cultures of the world, men are consistently more aggressive than women in most groups. From this evidence, our best guess is that both biological and cultural differences contribute to the greater incidence of violence in men.

People have been arguing for centuries about how men and women differ, so we don't expect to settle the issue here. As comedy writer Robert Orben said, "Nobody will ever win the battle of the sexes; there's just too much fraternizing with the enemy."

PART SIX

YOUR BRAIN IN ALTERED STATES

Do You Mind? Studying Consciousness

In Your Dreams: The Neuroscience of Sleep

A Pilgrimage: Spirituality

Forgetting Birthdays: Stroke

A Long, Strange Trip: Drugs and Alcohol

How Deep Is Your Brain? Therapies that Stimulate the Brain's Core

CHAPTER 26

Do You Mind? Studying Consciousness

The concept of free will presents an apparent paradox to anyone interested in the philosophy of how the brain works. On the one hand, your everyday experience—your desires, thoughts, emotions, and reactions—are all generated by the physical activity of your brain. Yet it is also true that the neurons and glia of your brain generate chemical changes, leading to electrical impulses and cell-to-cell communication. The implication, then, is that physical and chemical laws govern all your thoughts and actions—a proposition with which we wholeheartedly agree. Yet every day, we make choices and act upon the world around us. How can these facts be reconciled?

It is undeniable that brain injury can lead to changes in behavior. The nineteenth-century railway worker Phineas Gage was a responsible, hardworking man until a tamping rod blew upward through his lower jaw and out the top of his head. Amazingly, he survived. However, afterward he became a no-good, promiscuous layabout. His experience is the quintessential demonstration that our brains determine who we are.

Free will is a concept that is used to describe what an entire person does. If the behavior of an object can be predicted with mathematical precision, it doesn't have free will. Therefore simple objects such as atoms and particles don't have free will. According to one point of view, free will is ruled out by the idea that the output of our brains could somehow be predicted if we could know what was happening in every cell.

However, a more useful interpretation is that our intuitions fail us when we try to predict what a complex system is doing. No scientist has done a complete computer simulation of what even a single neuron does biochemically and electrically—let alone the hundred billion neurons in an actual brain. Predicting the details of what a whole brain will do is basically impossible. From a practical standpoint, that's a functional definition of freedom—and of free will. For a long time, neuroscientists were reluctant to examine such questions because many of them felt that ideas like free will and consciousness were so mysterious and undefinable that they would be impossible to study. But it turns out that some aspects of conscious experience, at least, can be addressed experimentally.

It's hard to study individual subjective experiences, the ones you might have wondered about in those late-night conversations in school. What is it in brain activity that produces the quality of "cold" or "blue," in the sense of what I feel and imagine you might feel? This seemingly simple question perplexes scientists, partly because it defines the question in terms of unmeasurable aspects of experience, what philosophers who study the mind call qualia.

By the same logic that has been used to discover what brain structures are involved in other

Did you know? The Dalai Lama, enlightenment, and brain surgery

Our fascination with the brain's influence on moral behavior is shared by the Dalai Lama, who made a speech to the annual Society for Neuroscience meeting in 2005. Sam asked His Holiness whether, if neuroscience research someday could allow people to reach enlightenment by artificial means, such as drugs or surgery, he would be in favor of the treatment. His answer surprised us.

He said that if such a treatment had been available, it would have saved him time spent in meditation, freeing him to do more good works. He even pointed at his own head, saying that if bad thoughts could be stopped by removing a brain region, he wanted to "cut it out! cut it out!" His homespun English and stabbing motions were unforgettable and would have been more disturbing coming from someone not dressed in the robes of a holy man.

However, he felt that such a treatment would only be acceptable if it left one's critical faculties intact. We were relieved to hear this, since it rules out the prefrontal lobotomy, a neurosurgical treatment invented by Egas Moniz and popularized with great enthusiasm in the mid-twentieth century by the American psychiatrist Walter Freeman. Prefrontal lobotomy is a radical procedure in which prefrontal lobes are disconnected from the rest of the brain. It became popular in mental hospitals, primarily as a means of controlling troublesome patients. The surgery did remove violent and antisocial impulses, but it also removed many functions that we associate with mental existence, such as goal-directed action planning, motivation, and complex reasoning. Thankfully, lobotomy has been largely abandoned as a surgical treatment.

Did you know? **Can brain scanners read your mind?**

Activity patterns in the brain are fantastically complex. At any moment, many millions of neurons in your brain are generating electrical impulses. Reading what is happening in all these neurons at once is beyond the capability of any current technology. Even with such a recording, converting the measurements to an interpretation of specific thoughts is the stuff of science fiction, and it may never happen.

On the other hand, simpler feats are possible. For instance, using functional brain imaging, strong emotional responses can be seen as increased activity in the amygdala (see chapter 16). Scientists can even use brain activity patterns to tell which of two competing images a subject consciously notices. One set is shown to the left eye, and another to the right eye, causing subjects' conscious perception to switch back and forth between the two pictures. Researchers can identify patterns of activity associated with a subject's awareness of either the left-eye or the right-eye stimulus. In this way, they can predict the consciously experienced stimulus—after observing the subject's response to hundreds of presentations of the images. Attempts to design a brain scanner that would detect lies run into similar problems, as they need to be calibrated to an individual's brain activity while he tells the truth or tells a lie. This may be a bit too much cooperation to expect from someone whose reward for helping you might be a prison sentence.

So if you're worried about having your mind spied upon, take comfort in knowing that, to the extent it's possible, it requires you to lie very still in a million-dollar scanner for many minutes and for the spy to be satisfied with knowing whether you are noticing things with your left eye or your right eye. In other words, you don't need a tinfoil hat to protect yourself—but keep up that poker face.

mental phenomena (such as vision), a pattern of brain activity that is uniquely associated with the conscious perception of sensory stimuli would be a signature of awareness. If scientists can define activity that occurs only when you notice a stimulus—and never at any other time—then they can legitimately claim to be studying brain activity that is related to awareness.

In one experiment, scientists presented subjects with two sets of pictures in quick succession and asked the subjects to detect some feature of the first set. Concentrating on the first set of pictures made it hard for subjects to detect a particular feature in the second set, a phenomenon known as attentional blindness. Some brain regions were activated every time, whether the subjects reported perceiving the second stimulus or not. These areas included the primary visual cortex, which is the first stop for visual information in the cerebral cortex. However, other regions were activated only on the repetitions when subjects reported that they could see the second stimulus. This experiment shows that visual stimuli can activate a surprisingly large number of brain regions without entering into conscious awareness, suggesting that conscious awareness is like a spotlight that focuses on specific stimuli and ignores others.

Even though conscious awareness only extends to a fraction of incoming stimuli, more information is available for your brain's use. People with a medical condition called blindsight have normal eyes, yet are unable to report any details in part, and sometimes all, of the world around them. They are, for most purposes, partially blind. Yet when asked the direction of a light source in the direction of their blindness, they often point in the correct direction, though they believe themselves to be guessing. How can this be?

Blindsighted people have no functioning primary visual cortex, through which visual information must pass to get to the rest of the cerebral cortex. Because of this damage, they are unable to consciously perceive visual information. However, sensory information goes to other places in the brain. As you may recall from chapter 3, visual information comes from the retina to the thalamus and is transmitted onward to the cortex. The retina also connects directly to the superior colliculus, a structure found in nearly all vertebrates. Visual information in this more ancestral brain region is not consciously perceived but can still guide actions such as pointing or moving the eyes.

The information that is available without our being aware of it can be quite complex. Scientists at the University of Iowa found a way to measure the gap between hunch and recognition. People were asked to play a pretend gambling game in which they could choose cards from any of several decks. Each card gave

Did you know? My brain made me do it: Neuroscience and the law

A schoolteacher couldn't stop leering at his nurse. An intelligent and normally reasonable man, he had been acting very strangely and collecting child pornography. He had been apprehended after making sexual advances toward his stepdaughter. Though he knew these things were wrong, he couldn't stop himself. He told the doctors he was afraid that he would rape his landlady. And he had a terrible headache.

A brain scan revealed a large tumor pushing on the front of his brain, near his orbitofrontal cortex, a structure involved in regulating social behavior. After removal of the tumor, his sociopathic tendencies subsided, and he lost interest in pornography. Other annoying symptoms went away too, such as a tendency to urinate on himself.

Although most cases of sociopathy are not associated so clearly with brain damage, the teacher's case illustrates the possibility that criminal behavior can be linked to specific structural brain defects. Linking anatomy to behavior was first attempted in the nineteenth century by pioneering criminologist Cesare Lombroso, who failed because he focused on now-discredited measures such as head shape. However, well-controlled studies done since Lombroso's time have shown that violent criminals have a notably high incidence of head injuries in childhood and adolescence, especially to the front of the head. Brain imaging methods also now make it possible to detect examples of gross brain injury or damage (such as from the teacher's tumor) that can affect behavior.

The ability to associate criminal behavior with brain structure raises the possibility of a novel defense, namely that accused criminals are not responsible for their own acts. At one level, this suggestion makes no sense—morally speaking, we are our brains and cannot claim to have been duped or mistreated by them. But does our increased understanding of the brain tell us anything about how some criminals ought to be dealt with?

The law already has a category for those who are not mentally competent to understand the moral consequences of their acts. In cases such as the teacher's, one possibility would be to modify the standard of mental competence to require not only moral awareness but also the ability to act morally. This would fit with the old principle that people are responsible not for what they think but for what they do. We might also benefit from reexamining how we punish criminals. Two goals of punishment are moral retribution for a crime and deterrence to others. But the teacher already knew that his acts were wrong, and people with his type of injury would not be deterred even by certain punishment. Indeed, this view has legal precedent: the U.S. Supreme Court ruled in 2002 that execution of a mentally retarded person was inhumane.

A new issue raised by neuroscience is technological. The state of mind of a person can be changed, by surgical removal of a tumor, for instance. Under these circumstances, the person punished may be different from the person who committed the crime. According to criminal law, someone who plans a crime in advance is said to have committed a premeditated act and is subject to more severe penalties than one who acted in the heat of the moment. Legal precedent therefore exists for the idea that people may not be fully responsible for their acts. Perhaps in the future, those with brain injuries, like unpremeditated criminals, will pay some appropriate but lesser penalty and also be required to receive treatment.

As neuroscience advances, associations between brain structure and function will certainly expand. One point of view asserts that punishment must take such new science into account. Is life imprisonment the most effective means of punishing a fifteen-year-old whose prefrontal brain structures are not yet done developing? Is repair of a criminal's brain preferable to punishment? The question of fixing a defective brain is particularly fraught with moral difficulty since it involves changing the very identity of a person. Perhaps the Dalai Lama's criterion of leaving critical faculties intact would come into play. Such questions of "neurolaw" cast old questions of moral behavior in a new light. In the words of cognitive neuroscientists Jonathan Cohen and Joshua Greene, "If neuroscience can change [moral] intuitions, then neuroscience can change the law."

instructions to increase or decrease their bankroll. Without the participants' knowledge, some decks of cards were stacked against them: these decks provided big wins but even bigger losses, for a net loss, while other decks provided small wins and even smaller losses, for a net gain. After losing repeatedly, subjects began to choose the more favorable decks but were unable to say why until after much further play.

Some of the early reactions to playing a losing game are seen in the orbitofrontal cortex, which we introduced in chapter 16. Patients with damage to this region, which lies above and around the eye socket, don't ever improve their performance in this game—or even show stressful responses to losing, such as developing sweaty skin. The evidence suggests that this brain area can detect bad events before we are consciously aware of a problem. Processing in the orbitofrontal cortex could thus be involved in the experience of having a bad feeling about something.

Lack of awareness can even extend to one's own actions. In the 1980s, a California research team asked people to tap a finger, at a time of their own choosing, and note the time of their decision by checking a clock. Brain regions responsible for triggering movements started generating activity half a second before any movement was made. However, subjects only reported awareness of their decision a few tenths of a second later, shortly before the movement began.

This finding contradicts our everyday idea of free will. The conscious decision to take action, an event that we associate with free will, comes only after the stirrings of the action have already been initiated in the brain. The only part of conscious awareness that preceded the movement occurred when subjects were asked to stop a movement that other parts of the brain had already initiated. In some sense, this is not free will, but a veto: free won't.

Is the feeling of intention caused by the brain's motor preparatory activity? Quite possibly. However, it appears that our awareness of our own actions can sometimes dawn after the moment when a decision is made. The net effect is that our brains produce our actions, but part of the decision-making process is complete before we are able to report it. In that sense, we are doers, not talkers.

CHAPTER 27

In Your Dreams: The Neuroscience of Sleep

o one is sure why sleep is so important to life. Almost all animals sleep—including insects, crustaceans, and mollusks—and sleep deprivation can be fatal. Most theories of sleep suggest that it's important for the brain. As animals have diversified, and their brains have become more complex, sleep has likewise become more complicated, developing from a single stage to multiple stages.

Across many species, sleep decreases heart, muscle, and brain activity, but leaves animals able to wake up if they are prodded hard enough. Most animals sleep at night, which makes sense because it's hard to see (or be seen) in the dark. Sleep allows animals to conserve energy and to match their own activity with periods of warmth and light.

Whatever sleep's function may be, it takes a powerful counteradvantage for a species to forgo sleep entirely. Of the few animals that never sleep, most are fish that must swim to stay alive, such as skipjack tuna and some sharks, which get enough oxygen only if water runs through their gills at a high rate. A similar problem is faced by dolphins, which are air-breathing mammals that have to surface often; they do this by sleeping with only half of their brain at a time so that they can keep moving. Other nonsleeping animals include cave-dwelling fish and a few mostly stationary frogs, of which it would be reasonable to ask the converse question: do they ever really wake up?

In lower vertebrates, sleep consists of a continuous rhythm of low brain activity. In reptiles, electroencephalographic (EEG) recordings during sleep show a slow rhythm in the form of spiky events, suggesting that many neurons are active in synchrony. These slow-wave spikes are reminiscent of slow-wave sleep, the deepest stage of sleep in people.

When birds and mammals arrived on the evolutionary scene, a new type of sleep arose:

rapid eye movement (REM) sleep. At the same time, non-REM sleep began to include intermediate stages in addition to slow-wave sleep. REM sleep is defined by the eye movements themselves (which you can see if you watch a sleeping person) and the electrical signature of cortical brain activity. This activity has a spiky quality that resembles awake activity, which earned REM sleep its other name, paradoxical sleep, because the brain's activity during REM sleep is not very sleeplike.

REM sleep is when nearly all dreams occur, especially the vivid ones, in humans and other mammals. Sleeping dogs, cats, and horses make sounds and fidgeting movements during sleep. Dreamers are prevented from triggering active movements, though, because commands from the cerebral cortex to drive movement are blocked by an inhibitory center in the brainstem that is activated during sleep. Inhibition from the cerebral cortex prevents us from acting out our dreams and probably accounts for the feeling of paralysis that is often reported during dreams, especially frightening ones. Experts believe its malfunction to be a likely cause of sleepwalking, and also suggest that it might be a cause of bedwetting by children. The inhibitory center can be removed surgically; after such an operation, cats arch their backs and engage in mock combat during REM sleep, suggesting that fights are a common component of cat dreams.

Whether REM sleep and dreaming have a biological function is hotly debated. One of sleep's functions may be to "consolidate" memories. Long-term storage of memory seems to undergo a conversion of some kind over weeks to months, as our memories of facts, events, and experiences are gradually transferred from an initial storage place in the hippocampus to the cortex. At the same time, memories of specific episodes are incorporated into more general knowledge known as semantic memory, in which people remember facts without knowing how they were learned.

A day's experiences are almost never the subject of dreams the same night but instead are incorporated into dreams only after a delay of a few days or longer. Perhaps this is because sleep helps us process them. When sleep is interrupted, some kinds of memories are slower to consolidate. The critical part of sleep for consolidating memories has been variously suggested to be slow-wave sleep or REM sleep; deprivation of either stage has some effect on memory reconsolidation, though most of the evidence (and research) has focused on REM sleep.

One reason that it has been difficult to study sleep's connection to memory is that sleep deprivation damages the brain and body. Sleep deprivation induces a stress response in which the hormone cortisol is secreted. It takes about four weeks of sleep deprivation to kill a rat, and about two weeks to kill a fruit fly. The longest bout of known wakefulness for a human is eleven days. This feat, which is recorded in the *Guinness Book of World Records*, is likely to stand because the book has closed this category due to the health risks. After a few days of sleep deprivation, humans begin to hallucinate. At such stressful moments, hormones like cortisol are released,

Did you know? Wake up, little Susie: Narcolepsy and modafinil

Narcolepsy is a disorder in which sufferers inexplicably fall asleep at all times of the day. This can happen not only during inactivity, but also at exciting moments. The disorder has been studied in a colony of narcoleptic dogs living at Stanford University. Playing with one of these dogs proceeds normally until the dog gets too excited, at which point it falls asleep. Both human and nonhuman sufferers of narcolepsy lack a particular type of the neurotransmitter peptide orexin. Orexins act on receptors in the hypothalamus, a command center for the regulation of sleep, aggression, sexual behavior, and other core activities.

Treatments for narcolepsy have not yet taken advantage of the discovery of orexins. Instead, most treatments stimulate the nervous system by influencing the action of monoamines, a large category of neurotransmitters that includes serotonin, dopamine, and noradrenaline. The drugs used for this purpose include certain antidepressants and stimulants such as amphetamine and methamphetamine. The problems associated with these drugs include side effects such as dizziness or, in the case of amphetamine and methamphetamine, the potential for addiction. Amphetamine can promote wakefulness at lower doses than those that lead to motor activation, suggesting that amphetamine's effects on waking behavior could potentially be separated from its other effects.

One drug that seems to induce wakefulness without affecting motor activity is modafinil (sold in the U.S., U.K., and other countries as Provigil), a drug that has become popular for the treatment of narcolepsy. Modafinil and amphetamine both enhance wakefulness in normal people and narcoleptics; neither has any effect on wakefulness in mice that are missing a molecule that transports dopamine out of synaptic spaces. This finding suggests that wakefulness is tied intimately to the brain's dopamine signaling system.

One of modafinil's applications is to enhance wakefulness and reduce risk in long-shift workers. In a U.S. Air Force study, modafinil was almost as effective as Dexedrine (an amphetamine) in enhancing performance during forty-hour shifts. The pilots showed increased alertness, confidence, and performance on simulated flight maneuvers. If modafinil is really not addictive, it is likely to gain in popularity in both narcoleptics and people who must work long hours.

Did you know? **Why are yawns contagious?**

Although we associate yawning with sleepiness and boredom, its function appears to be to wake us up. Yawns cause a massive expansion of the pharynx and larynx, allowing large amounts of air to pass into the lungs; oxygen then enters the blood, increasing alertness. Yawning is found in a wide variety of vertebrates, including all mammals and perhaps even birds, and can be observed in human fetuses after just twelve weeks of gestation. In nonhuman primates, yawning is associated with tense situations and potential threats. Functionally, yawns can be thought of as your body's attempt to reach a full level of alertness in situations that require it.

Yawns are contagious, as anyone who has attempted to teach a roomful of bored students knows. The reason for this contagion is not known, though it might be advantageous to allow individuals to quickly transmit to one another a need for increased arousal. A video of yawning also increases the frequency of yawning in chimpanzees and in monkeys.

Yawning is not contagious in nonprimate mammals. However, the ability to recognize a yawn may be fairly general: dogs yawn in response to stressful situations and are thought to use yawning to calm others. You can sometimes calm your dog by yawning.

The ability to yawn is buried in the brainstem. Some tetraplegics with tumors in their pons, which block the transmission of cortical movement commands so that they cannot open their mouths, can still yawn involuntarily. In these patients, the only place in the brain that can initiate a yawn is a group of neurons in the midbrain that relays movement commands from the brain to facial muscles. Some researchers believe that yawns may begin in these neurons. Yawning can even occur in people in a vegetative coma.

A particular oddity of having yawning mechanisms in a place as tightly packed as the brainstem is that signals can unexpectedly leak from one region to another. For instance, one side effect of Prozac is that in some women, yawning can trigger clitoral engorgement and orgasm, an accidental connection that (for a lucky few) would make boring situations far more interesting.

Seeing, hearing, or even thinking or reading about a yawn is enough to trigger one's own yawning circuitry. You may be attempting to suppress a yawn as you read this, as we did while writing it. (We don't take it personally.) Seeing a yawn induces activity in areas of the cortex that are activated by other visual stimuli and social cues. Although we have outlined why yawns would be contagious—the advantage of sharing an alert signal—we don't know exactly what happens in the brain to spread the contagion.

and these stress hormones are known to impair learning. Sleep deprivation's negative effect on memory can't be explained entirely by stress, though, as sleep deprivation still blocks memory consolidation in animals after their adrenal glands have been removed to keep them from releasing stress hormones.

Why would sleep be important for memory consolidation? One possibility is that changes in the strength of connections between neurons (synaptic plasticity) are driven by neural activity, which can occur whether an animal is awake or asleep. If neural activity from a remembered episode were replayed during sleep, it might facilitate memory consolidation in this way. Indeed, some patterns of waking neural activity are played back during sleep on a remarkably precise timescale, exact to thousandths of a second. One activity requiring precise sequencing of neural firing is the production of sounds, such as speech or birdsong. When a bird sings, specific sets of neurons in the bird's brain fire in an order that is linked closely to the sequence of sounds in the song. These neurons are responsible for generating precisely controlled changes in muscle tension that control the bird's sound-producing organ, thereby generating the same song every time. Researchers monitored these neurons while the birds slept and found that the same patterns were generated during sleep. In a sense, it appears that birds dream about singing their songs.

Non-REM sleep may also involve the playback of waking experience. As a rat runs through a maze, so-called place cells in its hippocampus fire in an order corresponding to the sequence of locations that the rat passes through. When the rat is asleep, the same place cells fire again in the same order. This replay occurs during slow-wave sleep, when dreaming is very rare in humans. The replayed snippets are typically a few seconds long, suggesting that rats replay moments in the maze, not necessarily the whole experience.

Synapses in different brain regions

follow different rules for the conditions under which plasticity can occur, and these differences may relate to the phases of sleep. For instance, in the hippocampus, where initial spatial and episodic memories are thought to be formed, changes in synaptic strengths require the theta rhythm, a pattern of about eight neural spikes per second that occurs only in awake animals during exploratory behaviors such as walking—and in REM sleep. For this reason, scientists associate memory consolidation with REM sleep.

The idea that sleep is important for reconsolidating and redistributing memories provides an alternative to Freud's view that dreams express unconscious desires. This piece of psychoanalytic folklore does not have any experimental proof, but is likely to have its roots in the observation that dreams often incorporate the daily concerns of the dreamer, combined with seemingly random or senseless events. The existence of trains of thought in dreams and a degree of plot suggests that your cortex has some ability to construct a coherent story from what it is given—though this may simply reflect the action of the "interpreter" discussed in chapter 1. In that respect, dreams may constitute a means of sampling what's lying around in your head. When we talk about our dreams, we focus on the extent to which our dreams can be made coherent: waking up in class naked, sailing a ship, rolling a big rock. But what if the random aspects of dreaming are an essential feature? What if randomly sampling the brain's contents as we sleep is a means for transferring our memories to a more permanent place? Resampling could even be used to correct wrong memories that need to be erased. Weird dreams may be the price, or perhaps an unintended benefit, of the mechanisms that our brains use to remember the events of our lives.

CHAPTER 28

A PILGRIMAGE: SPIRITUALITY

How religion is rooted in our biology has been a popular topic for recent books, especially among atheists who are convinced that religious beliefs are irrational. Prime examples are *The God Delusion* by biologist and bomb thrower Richard Dawkins and *Breaking the Spell: Religion as a Natural Phenomenon* by philosopher Daniel Dennett. Considering how little is known about the neuroscience of religion, it seems premature to claim that biologists have the issue all worked out.

Anthropologists have expressed a more positive view of religion: that it was a powerful early instrument of group social bonding, which may have provided a survival advantage for religion itself and for humans who shared the beliefs. Let's start by reminding ourselves that organized religion is a remarkable achievement, one of the most sophisticated cultural phenomena in existence. Consider the basic elements of most religions: believers have elaborate cognitive representations of a supernatural force that cannot be seen. We plead with the force to reduce harm, bring about justice, or provide moral structure. We furthermore create an understanding with our fellow humans that this force sets the same standards of morals, social norms, and religious rituals for all of us. It's a complicated business, unique to us among all living beings.

What can neuroscience contribute to our understanding of religion? In one sense, nothing: the satisfaction derived from religion is unlikely to be changed much by knowing how the brain gives rise to beliefs. Just as you can use words profitably for a lifetime without understanding formal grammar, people can benefit from religious belief—and for that matter, many other systems analyzed in this book—without understanding its basis in the brain. Still, if you've come this far, you might be curious.

Two brain capabilities are particularly important in the formation and transmission of religious belief. Many animals probably have some form of the first trait: the search for causes and effects. The second trait, social reasoning, is unusually highly developed in humans. One of the core skills of the human brain is the ability to reason about people and motives—what scientists call a theory of mind.

The combination of these abilities has generated key features of mental function that are part of religious belief: our ability to make causal inferences and abstractions, and to infer unseen intentions, whether they're the intentions of a deity or some other entity. Neural mechanisms that favor the formation of religious beliefs are also likely to favor the formation of organized belief movements of other kinds, including political parties, Harry Potter fan clubs—and militant atheism.

What kind of god would it be who only pushed the world from the outside?
—Johann Wolfgang von Goethe

Most religions seek causes for events in the world. These explanations often take the form of actions performed by a thinking entity. For example, small children either explicitly or implicitly assign motives to inanimate objects. Developmental psychologists find that small children think a ball rolls because it wants to. This way of thinking is so natural to us that we do not hesitate to think of everyday objects as having personalities. We often assign cars or other machines personalities and even names. Teakettles whistle cheerily, and storms rage. It seems natural, then, that early humans might have applied such reasoning to the events of the natural world. This kind of reasoning is seen in animist religions, which attribute a spirit to living and nonliving objects.

Applying the metaphor of conscious agency to natural events becomes something new when combined with our intensely social nature. We dedicate considerable mental resources to understanding others' motivations and points of view. The growing complexity of a child's view of motivation can be seen in play, which starts with simple sensory activation but quickly blossoms into first-order pretense ("I'm a wagon!") to baroque role-playing ("Okay, now you pretend to be the child, and I will pretend that I'm the teacher and the class is making too much noise").

The attribution of imaginary motives to oneself and to others requires a theory of mind. This ability allows children to engage in fictional play, like pretending that a toy soldier can fight. As they acquire a theory of mind, children realize that others have motivations, which they can use in innocent ways, such as games of hide-and-seek, but also to more nefarious

ends, like misleading another person. At later stages, the sophistication of pretense becomes even more complex; children develop the ability to understand a staged drama. In chapter 24, we explained that people with autism have difficulty in understanding that others have motivations and desires, which has profound and sometimes disastrous effects on their dealings with the world. So the theory of mind is central to our sense of ourselves and of others.

Assessment of social scenarios requires activity in many cortical areas. One example is mirror neurons, which fire both when a monkey performs a task and when he sees another monkey do the same task (see chapter 24), suggesting that the monkey's brain understands that the two actions share something in common. In addition, social communication is impaired in monkeys with damage to the amygdala (see chapter 16), a brain structure intimately involved in deriving the emotional significance of objects and faces, and therefore critical in giving the brain access to knowledge of the mental states of others. All this brain machinery is likely to be involved in our attempts to explain things like natural events and complex relationships among nonhuman or inanimate objects.

Religious belief is made possible when the drive for causal explanations is combined with our brains' ability—and propensity—to provide advanced levels of social cognition. Together, these two abilities allow us to generate complex cultural ideas ranging from jaywalking to justice, redemption to the Resurrection. As we noted in chapter 3, complex social reasoning is related to cortical size. This strong relationship implies that social cognition requires some

Did you know? Meditation and the brain

The Dalai Lama says that when scientific discoveries come into conflict with Buddhist doctrine, the doctrine must give way. He also has a strong interest in exploring the neural mechanisms underlying meditation. Like many practitioners, he divides meditation into two categories: one focused on stilling the mind (stabilizing meditation) and the other on active cognitive processes of understanding (discursive meditation). Neuroscience's first pass at studying meditation focused on the first category. Brain activity in highly skilled practitioners of stabilizing Buddhist meditation was evaluated by a group of scientists, including one with a Ph.D. in molecular biology who has since joined the Shechen Monastery in Nepal as a disciple.

The group was able to draw eight long-term practitioners of Tibetan Buddhist meditation away from their normal practice (which is spending all day in meditative retreats). In the laboratory, the monks had electrodes placed on their heads to measure patterns of electrical activity. At first, the patterns were no different than those of volunteers meditating for the first time. The difference came when the monks were asked to generate a feeling of compassion not directed at any particular being, a state that is known as objectless meditation. Under this condition, the activity began varying in a coherent, rhythmic manner, suggesting that many neural structures were firing in synchrony with one another. The increase in the synchronized signal was largely at rates of twenty-five to forty times per second, a rhythm known as gamma-band oscillation. In some cases, the gamma rhythms in the monks' brain signals were the largest ever seen in people (except in pathological states like seizures). In contrast, naïve meditators couldn't generate much additional gamma rhythm at all.

How brains generate synchronization is not well understood, but gamma rhythms are greater during certain mental activities, such as attending closely to a sensory stimulus or during maintenance of working memory. This increased gamma-band rhythm may be a key component of the heightened awareness reported by monks. Are monks born with a natural ability to generate a lot of brain synchrony? Several types of rhythm seem to get stronger with experience in novices who learn meditation, suggesting that the capacity is at least partly trainable.

Brain scanning also identified regions that are active during discursive meditation (focused attention on a visualized image). Anterior cingulate and prefrontal areas of the cortex were very active, as they are when Carmelite nuns recall the feeling of mystical union with God. This work fits with the involvement of these regions in attention. It probably would also have been of interest to Pope John Paul II, who, in reference to science and Catholic doctrine, said that they were both true and compatible with one another because "truth cannot contradict truth."

serious information-processing horsepower. The brain arms race that rewarded our ability to cooperate with and outsmart our fellow beings has also set the stage for religious mental constructs. As a consequence, we can imagine a God, Yahweh, or Allah that is the cause of everything and judges us, yet who cannot be seen.

If theory of mind is a critical factor in the formation of religion, then animals that display some kind of theory of mind might be capable of religious belief. Can animals form a mental model of what others are thinking? In some species, the answer might be yes. For example, consider our friend Chris's dog Osa. Because of an injury, Osa was temporarily unable to climb stairs by herself, and Chris had to carry her up and down. This persisted for months, with her waiting at the top or bottom of the stairs to be carried. One day Chris came home at midday and was quietly puttering around in the kitchen. Osa came down the stairs, walking. Halfway down, she saw Chris and froze with a look that seemed to say, "I am so busted," which, of course, she was. Osa appeared to be acting on the assumption that if Chris knew she could walk down stairs, he would stop carrying her. This suggested to Chris that she could visualize what made Chris tick, at least when it came to schlepping pets up and down the stairs.

It's a gigantic leap to assign a theory of mind based on a single look from a dog. One could just as well say that the story demonstrates Chris's own theory of mind. However, in more systematic studies, dogs do appear to take into account other dogs' attentional states when trying to persuade them to participate in play, adjusting the signals they send according to what the other dog is doing.

Ethologists and anthropologists study the degree of sophistication of theory of mind by counting how many levels of intention can be imagined. Osa's desire to deceive sits at a relatively simple theory-of-mind inference. (*Chris thinks I can't walk down stairs.*) In religious belief, the level of reasoning is more complex. Multiple steps of inference are needed to follow the mutual motivations of multiple entities. A core necessity in religion is to make, at a minimum, a two-step inference: *God thinks* (step 1) *that I should worship him* (step 2). The details of most religions involve more steps of inference. To take Christianity as an example, having to keep straight what one wants alongside the desires of God, Jesus, the Holy Spirit, the teachings of the church, and one's fellow churchgoers gets very complicated.

Most apes and monkeys seem unlikely to be capable of multistep inferences about mental states, a minimum condition for religion. But observations of big-brained apes such as chimpanzees suggest that they can achieve at least Osa's level of inference. For instance, a subordinate chimp will prefer to go after a piece of fruit that cannot be seen by a dominant chimp over one that is visible to the dominant chimp. Similarly, if you appear unwilling to give a grape to a chimp, it will lose interest. If you show the same chimp that you're willing but unable

Did you know? **The neuroscience of visions**

Mountains are important in the three major monotheistic religions practiced today: Judaism, Christianity, and Islam. All three involve special visions that occurred at great heights. Moses encountered a voice emanating from a burning bush on Mount Sinai. Jesus's followers witnessed the Transfiguration on what was probably Mount Hermon, and Muhammad was visited by an angel on Mount Hira (Jabal an-Nour). These visions are but three examples of a broader category of mystical experience. Yet another prominent example is the appearance of the Virgin Mary to Juan Diego as he ran across Tepeyac Hill in Guadalupe, Mexico. Commonly reported spiritual experiences include feeling and hearing a presence, seeing a figure, seeing lights (sometimes emanating from a person), and being afraid. Curiously, very similar phenomena are reported by a group generally not thought to be very mystical: mountain climbers. Could it be something about the mountains?

Mountaineers have long known to watch for the dangers of thin air. Acute mountain sickness occurs above altitudes of twenty-five hundred meters (about eight thousand feet). Many of the effects are attributable to the reduced supply of oxygen to the brain. Reaction times are measurably reduced at altitudes as low as fifteen hundred meters (about five thousand feet). At twenty-five hundred meters or higher, some mountaineers report perceiving unseen companions, seeing light emanating from themselves or others, seeing a second body like their own, and suddenly feeling emotions like fear.

Oxygen deprivation is likely to interfere with activity in neural structures in and near the temporal and parietal lobes of the cortex. These brain regions are active in visual and face processing, and in emotional events. An extreme case of disturbed function is an epileptic seizure. Temporal lobe seizures often result in intense religious experiences, including feeling the presence of God, feeling that one is in heaven, and seeing emanations of light. Temporal lobe seizures are triggered more easily under conditions that elevate endorphins, such as high stress. The exertion of climbing a mountain would certainly be a source of stress, and religious visions often occur under stressful conditions. Indeed, as a general rule, visions are associated not only with mountains but with other remote areas where environmental conditions are extreme, such as deserts. Seizures are thought to have caused religious visions in Saint Teresa of Ávila and Saint Thérèse of Lisieux, and may have triggered conversions of previously nonreligious people, including the apostle Paul on his way to Damascus and Joseph Smith, the founder of the Church of Latter-day Saints.

to give it the grape, it will wait longer. Chimpanzees make these inferences with a brain that is less than one-third the weight of ours. The jury is out on whether chimpanzees can form religious beliefs. In one behavior that is very suggestive, during thunderstorms some chimps sway around with their hair standing on end, an act that some people have interpreted as resembling a dance. Are they superstitious? Or just afraid? At this point, since the evidence for chimpanzees having a theory of mind at all is so recent, we can only wait for more information.

A final element in religion is the passing on of teachings and traditions. Such continuity requires language, which permits accumulated ideas to be modified, allowing doctrine and dogma to be communicated from generation to generation. For now, humans seem to be alone in having the basic mental tools—a theory of mind and language—to generate organized religion. But we may not have always been the only ones with this gift. Before our species made the leap to religious belief some tens of thousands of years ago, with our ritual burials and cave art symbolism, Neanderthals, another branch of the *Homo* lineage, may have done so as long as one hundred thousand years ago.

Our capacity for language allows our search for causes and effects to take on a new dimension in the form of narrative. Human beings are storytelling, narrating animals, and as such have developed complex explanations for a wide variety of daily experiences and problems of existence. In his book *Moral, Believing Animals*, sociologist Christian Smith discusses human belief systems as a general phenomenon in which the world is placed in a coherent conceptual framework, a story that gives meaning to daily experience.

The search for explanation in different contexts is a central feature of many belief systems. Examples of explanatory narratives include an understanding of historical events (political science), natural phenomena (science), and social dynamics (psychology and sociology). In this way, religion is an example of yet another narrative, one that looks for meaning in the experience of living—spirituality. Although all these forms of narrative use different rules and have very different goals, any eventual neuroscientific explanation of how we form narratives is likely to address them in similar ways.

Once explanatory ideas take root, the sky's the limit on what a conceptual structure can explain or recommend. Why should we not harm our neighbors? Where did Grandma go when she died? Who made the world? When we encounter unbelievers, should we attempt to kill them or convert them?

Of course, asking or answering these questions does not require belief in a god. In one episode of the animated television show *South Park*, Cartman travels to a hypothetical future in which three factions are engaged in a bitter battle for world dominance. They share in common a venerated founder and much doctrine, but a small difference has led them to fight to the death. Their doctrine? Atheism. Their founder? None other than Richard Dawkins. Speaking as neuroscientists, we find the most unrealistic aspect of this show to be not the warring factions of atheists, but the fact that one of the factions is composed of . . . sea otters. Real otters aren't likely to be able to keep track of enough actors to form a dogmatic belief system. Anyway, we hope not.

CHAPTER 29

FORGETTING BIRTHDAYS: STROKE

On a winter morning in 2002, Sam phones his mother in California. Today is his brother Ed's birthday, and his Chinese-born parents are not sentimental about birthdays. Besides, it gives him a reason to call her, which she tells him doesn't happen enough.

"Today is Edward's birthday," he tells her. "Oh, really?" she replies. "What's the date today?" A red light in Sam's mind starts blinking, and his mother also becomes anxious. She knows that she is supposed to know her son's birthday. Fighting his own rising panic, Sam starts asking her other questions. "When is my birthday?" She can't remember. "Mom, when is your birthday?" She draws a blank. What about the message she left last week about going to Europe together? Nothing.

By this time, she has also realized that something is seriously wrong. She starts writing down the answers to all these questions, writing down everything. His father comes on the line. He's not exactly sure how long she has been like this, but he becomes convinced that she needs to get to the hospital. At age sixty-six, Sam's mother has had a stroke.

A stroke is an event in which blood flow to a brain region is disrupted, either when a blood vessel breaks (a bleeding, or hemorrhagic, stroke) or because it becomes blocked (a clotting, or thromboembolic, stroke). The great majority of strokes start from a thrombus, a clot that forms in a blood vessel that is hardened by arteriosclerosis or otherwise damaged. The thrombus can form in the brain itself or travel from elsewhere and get stuck in the brain, in which case it is called an embolism. In all strokes, a part of the brain is deprived of oxygen and glucose, which delivers energy throughout the body, and waste products can no longer be carried off. These events resemble what happens during a heart attack, when the flow of blood to the heart is stopped. For this reason, a stroke is sometimes called a "brain attack."

Strokes can happen to younger adults but are more common among older people. In the U.S., people age fifty-five and older have a one-in-five chance of having a stroke during their lifetimes. Among men, the risk is slightly lower, but is still one in six. Last year, about seven hundred thousand people in the U.S. had strokes of some kind. Nearly five million survivors of stroke are alive today.

Pound for pound, your brain uses more energy than any other organ of your body. All of this energy is carried to your brain by the blood. If the blood flow stops for any reason, it can stop the functioning of neurons almost instantly. Different parts of your brain take on different tasks. For this reason, the symptoms of stroke are specific to the area of the brain that has stopped functioning.

The most common places for a stroke are the cerebral hemispheres, because they are the largest part of your brain—about four-fifths of the total volume. The most common symptoms of stroke are loss of the ability to move a limb or loss of sensation in a part of the body. The cortex is also required for thinking, and so another common symptom is confusion. Yet another symptom is a sudden inability to speak or comprehend language.

The symptoms of stroke can also occur in other cases, known as transient ischemic attacks, but the strokelike symptoms of these episodes reverse in minutes. These events are not well understood, but they are probably caused by a loss of blood flow. Maybe a small clot forms, slowing blood flow by just a little, then dissolves.

In stroke, the initial symptoms persist. If the blood flow is stopped for more than a few minutes, neurons begin to die. Over the next hours, up to a day, damage gets progressively worse. By the end, many neurons have died. In one estimate, each minute of blocked blood flow destroys 1.9 million neurons, 14 billion synapses, and 12 kilometers (7.5 miles) of myelinated axons.

Can the effects of stroke be reversed? Currently, the answer is yes, but only in the first three hours. During this short window, if the victim is taken to an emergency room and diagnosed, it may be possible to give drugs that reopen clogged vessels or treat bleeding. After that, neurons are on their way to dying, and it's mostly too late to help. However, only a small fraction of stroke victims ever receive this treatment, the ones who are taken to the right emergency room, usually at large urban hospitals.

Four days later, Sam is with his parents in their doctor's office. By now, irreversible damage has been done, but he doesn't know this yet. He has heard about new stroke treatments, and he is hoping that something can be done. His parents came to the U.S. in the 1960s, and they have an attitude toward doctors that is common among older immigrants—they go back and forth between being intimidated and trusting everything the doctor says. Sam figures he'd better be there.

His parents' family doctor, who works at the local community health center, is a nice old fellow, also Chinese, also an immigrant, not far from his parents' age. For these reasons, they like him. He comes in, a bit harried. He is friendly, but he appears to have made his mind up about Sam's mother. Outside the office are many patients, some in other examination rooms, waiting for him.

Sam tries to persuade the doctor that his mother has had a stroke. The doctor is skeptical because she also had some gradual memory decline before this event, a common sign of Alzheimer's disease. But this diagnosis doesn't make sense; her recent memory loss is large and sudden. She has diabetes, which is a risk factor for microstrokes and stroke in general. This could explain both the gradual and sudden declines. Still he resists the idea, perhaps because stroke would often cause a sensory, movement, or language problem in addition to the abrupt memory loss.

Together they look at the MRI scan report. It comes up mostly normal, but a phrase leaps out at Sam: "anomalous low contrast focus in the anterior left thalamus, 4 mm wide." This means that the picture shows a little spot of dead tissue, a lesion, deep in her brain. This is it. This is the damage. Her thalamus has been damaged by a tiny blood clot lodged in a blood vessel.

The doctor is not convinced. He says, "This lesion is so small, smaller than the nail of your little finger." He finished medical school nearly forty years ago. He might not even have taken a neurology class. At some schools, it's not required. The thalamus itself is less than an inch long. It transmits information from one part of the brain to another, especially sensory information to the cerebral cortex. But it also communicates with parts of the brain involved in memory. In the thalamus, four millimeters is a big lesion. Eventually the doctor agrees to refer Sam's mother to a neurologist. Fifteen minutes later, he is off to his next patient.

Since her stroke, Sam's mother has been having trouble learning about new facts and events. A related form of memory is spatial navigation—the memory skill you use to get to your favorite neighborhood coffee shop even before you've had the benefit of your morning coffee. These forms of memory require structures located on the sides of the brain and in its core, in regions known as the temporal lobe system (see chapter 23).

The role of the thalamus in memory is relatively mysterious, partly because it's composed of many different nuclei (clusters of neurons). Some of these nuclei transmit sensory and motor information. Others connect to assorted brain regions involved in other functions. We don't know what many of these nuclei do. In the laboratory, the way we find out is to damage a nucleus and see what goes wrong, or to record electrical activity. We can also trace the wiring, the way that you might trace a cable from the back of your stereo. In humans, it's unethical to deliberately damage parts of the temporal lobe or track the wiring inside the living brain. Therefore, stroke victims are a useful source of information. Useful for the student of the brain, that is—unfortunate for the patient.

The thalamus, a small brain component, is a less-common place for strokes to occur than the cortex, and a memory deficit after a thalamic stroke is unusual. This is partly because the thalamus is a gateway to all parts of the cerebral cortex, and only certain of these pathways are involved directly in memory.

With the specialist, Sam looks at a new set of MRI scans that reveal more detail than the ones taken at the community hospital. In this scan, his mother's brain shows two small spots, very close to one another, in her anterior left thalamus. It looks as if a sharpshooter had aimed a BB gun at a target. The doctor explains that the sharpness of these spots is strong evidence that a blood clot did indeed lodge in her brain. A stroke of the other kind, caused by bleeding, probably would have resulted in more widespread damage.

Sam's mother had two big risk factors. First, her father had heart disease and may have died of a stroke. This family history indicates that she may have inherited a predisposition to have a stroke. Second, she has diabetes mellitus, a condition of high blood sugar, which she has not been treating properly. For reasons that are not entirely known, untreated diabetes and high blood sugar increase the risk of stroke. One possible reason is that diabetics have impaired blood flow, which can increase the risk of clot formation.

The specialist gives her some basic neurological tests. One is the three-object test. He gives her three words—*blue, Paris, apple*—then changes the subject. Five minutes later, he asks for

Practical tip: **Warning signs of stroke—and what to do**

Detection: How can you tell if you are having a stroke? If you experience a sudden loss of feeling or movement in a particular part of your body, this may be a stroke or "brain attack." You may also experience a sudden inability to speak or recognize speech. If any of these events occur, it is essential to get to an emergency room right away.

Treatment: After a stroke occurs, immediate treatment, within three hours of the stroke, can help reduce the damage. Only some large hospitals have the ability to diagnose and treat stroke, so it's a good idea to identify the right hospital in advance. The type of treatment depends on whether the stroke is the more common kind, blockage of a blood vessel (ischemic) or the less common kind, bleeding (hemorrhagic). For clotting strokes, clot-busting drugs such as tissue plasminogen activator (tPA, also called Activase or alteplase) can help. For hemorrhagic strokes, tPA would worsen the damage. In hemorrhagic strokes, the treatment options are not as good, but can include drugs.

Prevention: Lifestyle changes can help prevent stroke. Smoking and excessive alcohol intake are major risk factors. Diets high in sugars and in saturated fats such as red meat and eggs are associated with stroke. Conversely, green vegetables; some fish, such as salmon, mackerel, and tuna; and the use of certain fats, such as canola, sunflower, or olive oil, in cooking can all lower your risk. Finally, regular exercise reduces the chance of stroke.

Major predictors of stroke, especially in people over fifty-five, include excessive weight, high blood pressure, and untreated diabetes. All of these predictors can be detected by routine physical examinations. A history of previous stroke or transient ischemic attacks is also an indicator of possible future stroke.

An additional way to prevent clotting strokes, the most common form, is the use of antiplatelet medications. The most commonly available one is aspirin, which in small doses reduces the risk of stroke and heart attack. Other antiplatelet drugs are also available that attack clotting mechanisms more strongly. However, antiplatelet drugs are not appropriate for some patients, including people with gastrointestinal bleeding.

To learn more about stroke, go to http://www.strokecenter.org.

the three words back. Nothing. However, she can do other things, like count backward by sevens: one hundred, ninety-three, eighty-six . . . She can touch her nose with her eyes closed. Many functions are fine—but not memory. She has also lost some memory of events before the stroke. She can't remember the terrorist attack of September 11, 2001—less than five months ago. Who could forget it? Now that's memory loss.

The doctor thinks her memory will improve somewhat over the next several years as her brain rewires itself to get around the new damage. However, a full recovery is not likely. In the meantime, there are new drugs that have some effect on memory loss, both in Alzheimer's disease and in stroke-induced memory loss. These drugs affect the neurotransmitter systems acetylcholine or glutamate. He prescribes one.

Over the next few years, Sam's mother's function improved somewhat. She eventually learned to pass the three-object test, so that the game of giving her three things to remember stopped fascinating the family. She could remember things for many days, like when Sam was coming to visit next, or what happened in the news the previous week. At the same time, her memory, which had once been prodigious, was still severely impaired. She used to run a brisk business selling and developing real estate, which required the continual recall of many facts, and going back to work remained impossible for the rest of her life.

CHAPTER 30

A Long, Strange Trip: Drugs and Alcohol

William S. Burroughs was fascinated with altered states of experience. A lifelong drug user, Burroughs wrote about his reactions to heroin, methadone, alcohol, cocaine, countless hallucinogens, and other drugs in books like *Junky*, *Naked Lunch*, and *The Yage Letters*.

Even so, Burroughs experienced only a small fraction of the hundreds of mind-altering substances in the world. Most of these drugs work by interfering with the actions of neurotransmitters. Some drugs mimic the action of a naturally occurring transmitter; others enhance or block the action of transmitters. You may recall from chapter 3 that some receptors respond to their transmitters by generating electrical signals that affect the likelihood that the neuron will fire a spike. Another type, called metabotropic receptors, generates chemical signals that affect the internal workings of the cell. Metabotropic receptors are frequent targets of mind-altering drugs. Their job is to modulate the functions of neurons or whole networks, often in subtle ways, making them essential in governing mood and personality.

The stars of this world are the monoamine neurotransmitters, which regulate mood, attention, sleep, and movement. The monoamines include dopamine, serotonin, adrenaline, and noradrenaline. These busy molecules are important in Parkinson's disease, Huntington's disease, depression, bipolar disorder, schizophrenia, headache, and sleep disorders.

Many mind-altering drugs interact with serotonin, which regulates sleep and mood. Serotonin interacts with more than a dozen receptors, each of which is found in a different subset of cells. Squirts of serotonin over here can make a neuron spike faster, over there make it more sensitive.

Did you know? **Ecstasy and Prozac**

Ecstasy and Prozac have very different uses: the first is a club drug, and the second is a treatment for depression. Surprisingly, both drugs have the same effect on the same molecular target. After serotonin is released, it is removed from the synapse by a transporter protein that sucks it into nearby neurons. Both Ecstasy and Prozac block the action of this transporter.

MDMA (methylenedioxymethamphetamine), better known as Ecstasy, was first synthesized in 1912. In the 1960s, MDMA was introduced in psychotherapy because it induces intense feelings of well-being, friendliness, and love for other people. For similar reasons, it became popular at nightclubs several decades later.

MDMA prunes back serotonin-secreting nerve terminals for a period lasting up to several months, though without killing the neurons. It may have some risk for addiction relating to its amphetamine-like structure, but the abuse potential is mitigated because the emotional effects of the drug diminish with repeated use. (Contrary to a myth, MDMA use does not deplete spinal fluid. This tale started from a study in the 1980s, in which MDMA users volunteered to give spinal fluid for analysis, and the rumor mill distorted the findings almost beyond recognition.) Ecstasy's effects begin soon after it is taken and last for many hours.

Prozac, on the other hand, requires repeated use over many weeks to take effect. Like Zoloft and Paxil, Prozac is a specific serotonin reuptake inhibitor, one of the most commonly prescribed types of drug. Although we know what these drugs do at a molecular level, exactly how they affect mood is not known. One possibility is that brain neurochemistry may adapt to the repeated administration of these drugs, for instance by making less serotonin to compensate for the extra serotonin hanging around at synapses.

An unresolved question is why a single dose of Prozac does not lead to Ecstasy-like effects. One possibility is that these drugs enter the brain at different rates. If Prozac enters the brain more slowly than Ecstasy, it might not give the same initial rush. Another possibility is that Ecstasy, which is structurally similar to amphetamine, can also block dopamine uptake, leading to effects similar to those of cocaine and amphetamine.

Because there are so many receptors for serotonin, it is possible to play with them in subtle and interesting ways.

Many hallucinogenic drugs are naturally occurring chemicals, such as those found in magic mushrooms and peyote, but the most precisely acting hallucinogen is the synthetic chemical

Did you know? Does marijuana cause lung cancer?

Everyone knows that tobacco causes cancer, whether it's smoked (lung cancer) or chewed (lip, tongue, cheek, and esophageal cancer). You might expect marijuana to pose a similar risk because both marijuana and tobacco smoke contain tar. By this reasoning, a marijuana joint might be about equivalent to an unfiltered cigarette. Most published studies on this topic have failed to exclude tobacco users from the test group, making it hard to know whether the cancers that occurred are attributable to tobacco or marijuana. Another error in these studies is the failure to distinguish among types of marijuana use (smoking a pipe or a joint, eating brownies, or smoking a water bong). So, as scientists like to say, the question needs more study. Volunteers?

lysergic acid diethylamide (LSD). LSD, or acid, is not addictive and causes no lasting organic damage to the brain. It binds very tightly to particular serotonin receptors, so doses of LSD are extremely small, typically between twenty-five and fifty micrograms—one ten-thousandth the weight of an aspirin tablet.

The tightness of LSD interactions is good for the physical safety of users. Basically, you can't overdose on LSD because it binds so specifically. Side effects occur because most drugs bind not only to their intended receptor, but also to other receptors, usually with lower strength. (Imagine if your front door key unlocked your neighbor's house some of the time.) In contrast, natural hallucinogens such as mushrooms contain many chemicals, which activate multiple receptors. Even without physical side effects, though, some acid trips can be upsetting, with long-lasting psychological effects. On rare occasions, LSD can cause psychosis, most often in users with an existing tendency toward mental illness.

Hallucinogens often produce powerful, consciousness-altering experiences. LSD brings out amazingly vivid imagery and appears to allow thoughts and perceptions that would otherwise be inaccessible. Poet Anne Waldman once described to us a trip in which she stood in front of a full-length mirror, seeing herself aging from a little girl to an old woman continuously. She saw herself at every stage of her life, separately and together all at once.

Another psychoactive substance that acts through metabotropic pathways is delta-9-tetrahydrocannabinol (THC), the active ingredient in marijuana. THC activates brain receptors that normally respond to cannabinoid neurotransmitters, which occur naturally all over the brain. THC reduces the likelihood that active neurons will release the neurotransmitters glutamate and GABA (gamma-aminobutyric acid, the most abundant inhibitory neurotransmitter in the brain) to excite or inhibit other neurons. In the normal brain, this depressed release is triggered by particular postsynaptic neurons, which secrete cannabinoids that are picked up by the presynaptic neuron. Taken as a drug, though, THC reduces the communication of many neurons non-selectively.

Another common drug, caffeine, has the opposite effect, enhancing transmission at many glutamate- and GABA-releasing synapses by increasing the likelihood of neurotransmitter release. Caffeine does this by blocking yet another metabotropic receptor, one whose normal job is to bind to the neurotransmitter adenosine. In this way, coffee is the antipot, as the drugs have opposing effects on brain function. Caffeine is a mild stimulant and a cognitive enhancer.

> If it weren't for the coffee, I'd have no identifiable personality whatsoever.
> —David Letterman

Another cognitive enhancer is nicotine, one of the most addictive drugs known, in vulnerable people, which acts on acetylcholine receptors in the brain. Nicotine addiction takes the form of intense cravings that lead to continued use even in the face of cancer risk. Smoking by pregnant women reduces birth weight and damages the brains of developing fetuses.

A major class of recreational drugs is the opiates, which include heroin, morphine, and many prescription painkillers (like OxyContin and Percocet). They act on the body's own pain-relief system, through receptors called opioid receptors, which are activated by neurotransmitters called endorphins. The greatest biological danger from opiate abuse is overdose, which can lead to respiratory failure and death.

The abuse of opiate-based painkillers can cause profound hearing loss. In 2001, right-wing radio personality Rush Limbaugh reported that he had lost most of his hearing. He later had an electronic device placed in his skull to restore it (see chapter 7). Although he claimed that his hearing loss was due to a rare autoimmune disease, it eventually emerged that he was an abuser of OxyContin. This provided a much more plausible explanation; opiate abusers often lose their cochlear hair cells for reasons that are unclear, though it is known that cochlear hair cells make opioid receptors.

Despite his opiate dependency, Burroughs lived to the age of eighty-three. In some sense, his long lifespan is not surprising. An opiate habit by itself is not life threatening, though withdrawal symptoms are very unpleasant. In later life, Burroughs maintained himself on steady levels of methadone, an opiate that prevents withdrawal symptoms but is slow-acting and therefore does not give the transient high, and consequent desensitization, that leads to a need for larger doses. As an experienced and wily user, Burroughs was able to function for many years.

A telling contrast is his son, William Jr., who also wrote about his experiences with drugs, but died of drug-induced liver failure at the age of thirty-three. The drug that killed him?

Did you know? **Hit me again: Addiction and the brain**

Some people just can't seem to stop. Drug use has enormous negative consequences in their lives, but they keep on taking their favorite drug. If you've ever wondered, "What is wrong with that person's brain?" you've got plenty of company. Neuroscientists have spent thousands of hours studying how drugs and addiction influence the brain.

Chronic drug use causes major changes in many brain areas. These areas include the brain's memory system, suggesting that powerful emotional memories or drug-taking triggers are involved in the development of addiction, as we know from the tendency of recovering addicts to relapse when confronted with drug-associated cues.

As we explain in this chapter, recreational drugs act on many different neurotransmitter systems, but they seem to converge on two areas that are part of the brain's reward system (see chapter 18). All addictive drugs cause the release of dopamine in the nucleus accumbens. Many also cause the release of endorphins and endocannabinoids in the nucleus accumbens as well as the ventral tegmental area.

Chronic drug use leads to a reduction in dopamine release. This change seems to cause reduced responses to natural rewards, such as food, sex, and social interactions, which involve some of the same brain areas. In nonhuman animals, repeated drug taking is associated with reduced functioning of prefrontal cortex neurons that project to the nucleus accumbens, which normally controls response inhibition and planning. Human addicts also show reduced prefrontal cortex activation in brain imaging studies.

A major problem with treating drug addiction is that responses to drugs and natural rewards overlap in the brain, making it difficult, for example, to target the desire for heroin without impairing the desire for food. Several drugs currently approved for the treatment of drug abuse are also under study as treatments for overeating, including rimonabant, which blocks cannabinoid receptors (see chapter 5). One way around this problem is to vaccinate people so that they produce antibodies against particular drugs, which prevent them from reaching the brain. A vaccine against cocaine is currently in clinical trials.

Amphetamine. Cocaine, amphetamine, and methamphetamine block the transport of dopamine. They are highly addictive and can cause widespread brain damage, particularly in developing fetuses (which are affected when drugs are taken by pregnant women).

All these drugs act by known pathways, though how they influence our behavior is not completely clear. But there is another common drug that is more of a mystery. It interferes with many elements of our biochemistry, and we still don't know exactly how it intoxicates us. Heavy use can lead to addiction, and in the long term, brain damage. Withdrawal symptoms brought on by sudden abstinence can be fatal. In most cases, it's legal. That drug is alcohol.

Until a few years ago, many scientists thought that alcohol led to intoxication by acting on the membranes that form the boundaries of cells, which are made mostly of fats. The idea was that if enough alcohol got into the membrane, these fats would move around more easily, interfering with the operation of receptors and ion channels.

Researchers now believe that alcohol has specific effects on neurotransmitter receptors that sit in the membrane. GABA's major target in the brain is the $GABA_A$ receptor, which produces electrical signals by allowing negatively charged ions to enter the cell, making neurons less likely to fire action potentials. Ethanol makes this channel stay open longer than it normally would, increasing the strength of this inhibitory signal, at a concentration similar to the one found in the blood of intoxicated people. (Alcohol also affects other ion channels, so intoxication may have multiple components.)

"When you drink, you're killing brain cells." How many times has this been said in bars

Practical tip: **Drinking and pregnancy**

Although alcohol in moderate doses does not kill mature neurons, it can have strong effects on developing neurons. Because nearly all neurons are formed and travel to their destinations before birth, the fetal brain is vulnerable to drinking during pregnancy.

Alcohol can kill newborn neurons, prevent their birth, and interfere with their migration from their birthplace to their eventual destination. In a fetus, even a brief elevation in blood alcohol is enough to cause some neurons to die. Two major components of fetal alcohol syndrome are a shrunken brain and a reduction in the number of neurons. Other factors that prevent neuron migration and survival are cocaine use or exposure to radiation.

around the world? The idea, firmly embedded in the culture and humor of drinking, rests on the mistaken presumption that if a lot of alcohol causes a lot of damage (it does), then moderate amounts of alcohol must cause some damage (not so).

Compared with teetotalers, heavy drinkers are likely to have shrunken brains, especially in the frontal lobes of the cortex, which is the seat of executive function. Magnetic resonance imaging was used to examine the fluid space that cushions the front of the brain from the skull in more than fourteen hundred Japanese people, ranging from abstainers to heavy drinkers. The skull does not change shape in adulthood, so expansion in this space indicates brain shrinkage. On average, heavy drinkers were more likely than nondrinkers to have brain shrinkage beyond that expected for their age. For instance, about 30 percent of abstainers in their fifties had brain shrinkage, while over 50 percent of heavy drinkers showed shrinkage. Changes were found in white matter, the axons that project from neurons to other parts of the brain, and gray matter, which contains neuronal cell bodies, dendrites, and the beginnings and endings of axons.

The reduction in gray matter is probably what started the idea that alcohol kills neurons, since an obvious explanation for shrinking brains would be neuron loss. However, this is not what happens. The cell bodies of neurons constitute only about one-sixth of the brain's total volume, while dendritic and axonal branches take up most of the space in gray matter. Indeed, no difference is seen between alcoholics and nonalcoholics in careful counts of neurons. (Of course, researchers do not count all fifty billion neurons. Instead, they sample the cortex at a number of locations and extrapolate the totals.) So what could account for the decrease in brain volume? In laboratory animals, chronic alcohol consumption leads to a reduction in the size of dendrites, which could yield decreases in volume without affecting neuron count.

The distinction between losing neurons and losing dendrites or axons is important. Loss of neurons would be very hard to make up, because in the cortex of adult brains, new neurons are generated at an extremely low rate, so low that some laboratories are unable to detect it at all. Meanwhile, shrunken cells, dendrites, and axons are capable of growth.

Does the brain recover when a human or animal gives up alcohol? After a few weeks, both brain volume and function begin to be restored. In animal experiments, cutting off the sauce restores the dendrites' complexity. In humans, alcoholics who give up drinking without relapses improve in cognition and a variety of other abilities, as well as in walking coordination. Human brains even show evidence of increased volume, which suggests that their brain cells reexpand, as has been seen in laboratory animals.

Though some of the effects of heavy drinking can be reversed, its consequences can nevertheless be quite serious. Drinking large quantities of alcohol over a long period of time is associated with many disorders, including high blood pressure and dementia. Although almost everyone's brain shrinks somewhat as they age, the shrinkage that occurs in heavy drinkers seems to be associated with serious cognitive and neurological deficits.

Furthermore, as we mentioned in chapter 1, years of heavy drinking can lead to a form of dementia called Korsakoff's syndrome, in which old memories are lost, and sufferers are unable to form new memories. In this syndrome, alcoholics develop a thiamine deficiency, which kills neurons in certain parts of the brain, including the anterior thalamus and mammillary bodies, which are connected to the hippocampus. These regions are part of the brain's system for storing new memories and eventually transferring them to long-term memory. In Korsakoff's patients, the loss of neurons—and of function—is irreversible.

A more relevant question for many of us is whether your brain is damaged by moderate consumption of alcohol. The answer is no. Many people assume that moderate drinking will have the effects of heavy consumption, only less severe. This is not always the case. Many processes that work to counteract damaging events can cope better with small events than with large ones. For example, the blood loss from a small cut is easily recoverable, but severe blood loss may be fatal.

The Japanese study we mentioned earlier showed that ingesting up to fifty grams of ethanol per day (three to four typical drinks of wine, beer, or liquor) has no measurable effect on brain structure. The consensus of many studies is that men can have up to three drinks a day, and women up to two drinks a day without adversely affecting brain structure or cognitive ability (except while you're drunk, of course). These numbers are handy because they mean a man and a woman together won't be harmed by five drinks, which is the amount of alcohol in a typical bottle of wine. A bottle of Pinot Noir per day per couple—sounds good to us.

The consumption of red wine may actually be beneficial. Drinking up to three or four glasses per day reduces the risk of dementia by a factor of two. As little as one glass three or four days a week can be beneficial, so the range of benign dosages seems to be fairly broad. Unlike hard liquor or beer, red wine decreases the risk of stroke, say several studies—including one from Bordeaux, France, where they know a thing or two about red wine.

Dementia can result from the cumulative effect of many small strokes, so it's likely that by reducing the risk of stroke, red wine consumption can preserve mental function. What we don't know is what is so special about red wine, and whether its alcohol content contributes to the beneficial effects. If the components of red wine that are responsible for this benefit are ever found, it may be possible to deliver them without the need to drink wine. This discovery would be useful—though a bit of a party pooper.

CHAPTER 31

How Deep Is Your Brain? Therapies that Stimulate the Brain's Core

Eighteenth-century Italian anatomist Luigi Galvani discovered that the nervous system uses electricity to convey signals. His assistant noticed that a frog's legs contracted violently when a nerve was touched by a metal scalpel. They subsequently found that small electrical sparks delivered to the leg were sufficient to generate contractions, a discovery that led to the modern understanding that nerves work by generating electrical impulses. Thanks to his discovery, Galvani's name has passed into popular consciousness: when an event suddenly arouses us to awareness or action, we are said to be galvanized.

Galvani's discovery eventually gave new hope to sufferers of a variety of neurological disorders, including Parkinson's disease and intractable depression. Stimulation deep in the brain's core can alleviate symptoms. Patients who receive deep-brain stimulation are galvanized, in the oldest sense of the word. The treatment can be quite effective, but we have little idea how it works.

Parkinson's disease strikes adults, usually in their fifties but sometimes earlier. Starting with a small tremor in voluntary movements, coordination gradually grows worse, and initiating a movement becomes harder. Late in the disease, patients develop muscular rigidity; even the smallest movements are slow and require a huge effort. Sufferers shuffle when they walk and often have their faces frozen in a mask. When Sam met a friend's wife who has Parkinson's disease, seconds passed before she was able to move. In the interim, the only clue to her intention was the focused look in her eyes and an increasing tremor in her hand, which grew to a marked swing as her desire to shake hands became apparent.

About 1.5 million people in the U.S. have Parkinson's disease, which affects about one in one hundred people over sixty-five. Famous sufferers include actor Michael J. Fox, boxer Muhammad Ali, Pope John Paul II, evangelist Billy Graham, and former U.S. attorney general Janet Reno. In some cases, such as Ali's, a contributing factor was a lifetime of small head injuries. But in general, the causes of Parkinson's disease are mostly unknown, as it is not very heritable.

The part of the brain most visibly affected by Parkinson's disease is the substantia nigra, a region deep in the brain that appears black in autopsies. The color comes from the neurotransmitter dopamine, which turns black when it oxidizes. In Parkinson's patients, these dopamine-producing cells die.

All treatments for Parkinson's disease focus on a network of areas in the brain's core that coordinate movement. The substantia nigra is just one of a group of neuron clusters, called the basal ganglia, nestled beneath the cortex. (The basal ganglia, which also include the globus pallidus and the subthalamic nucleus, communicate with one another and with other brain regions, such as the striatum.) The first surgical therapy for Parkinson's disease was to intentionally damage one of the basal ganglia structures. The idea that damage could be beneficial came from a chance discovery by neurosurgeons who accidentally ruptured a blood vessel (oops!) that supplies oxygen and glucose to parts of the thalamus, and found that their mistake had the unanticipated benefit of getting rid of the patient's tremor. The surgeons surmised that the death of some part of the affected tissue was responsible for the relief of the symptoms. This discovery was eventually turned into a strategy in which a small part of the basal ganglia complex is purposely burned away. This crude treatment, called a thalamotomy or pallidotomy, is sometimes effective, but did not become widespread since less than half of patients get any benefit. Even in the patients who do benefit, symptoms return after a few years.

Another development that made surgery less popular was the advent of a new idea for therapy: if dopaminergic neurons are dying, why not administer a drug that replaces dopamine? The best drug for this purpose turned out to be L-dopa, also known as levodopa, a chemical that can enter the brain, where it gets turned into dopamine. L-dopa and other drugs affecting the dopamine system are now the most popular therapies for Parkinson's disease.

Unfortunately, L-dopa only works up to a point. Like all neuromodulators, dopamine has multiple roles in brain function. For instance, schizophrenia is commonly treated with drugs that block dopamine receptors. Anti-schizophrenia drugs reduce psychotic delusions but often have the side effect of inducing muscle rigidity, shuffling gait, and masked facial expressions that look very much like Parkinson's disease. Conversely, L-dopa, which acts indiscriminately to strengthen dopamine's action everywhere, often leads to psychotic symptoms, such as hallucinations and delusions. As Parkinson's disease worsens, the benefit of drug therapy is limited because larger doses are more likely to trigger psychosis. Worse yet, L-dopa can have mixed positive and negative effects on movement, causing arms and legs to flail in sudden and unpredictable ways.

L-dopa's status as the best available therapy changed with a discovery made in 1986 by a French neurosurgeon who was performing a thalamotomy to correct a persistent tremor. As he worked, he was able to monitor the patient's movements and speech because the operation was done without general anesthesia. (This is possible because surgeons can get through the skin and skull with local anesthesia, and there are no pain receptors on the inside of the brain.) He used a small probe to deliver electrical shocks to help him find the place where he should make the lesion. At one location, he noticed that when he turned up the frequency in the electrode, his patient's tremor subsided. He later noted that the improvement he observed during the operation was just as good as the patient's performance after the thalamotomy.

This observation suggested that stimulation could somehow lead to an outcome similar to killing a bit of brain tissue. In the next few years, he tested this idea on patient after patient, giving them implants and battery packs that they could carry with them, allowing them to receive jolts all day. The benefits to these patients were striking. Patients who used to need caregivers once again were able to live independently. Some of them, previously on doses of L-dopa high enough to trigger unacceptable side effects, now needed far less medication and sometimes none at all. The therapy improved nearly all types of movement.

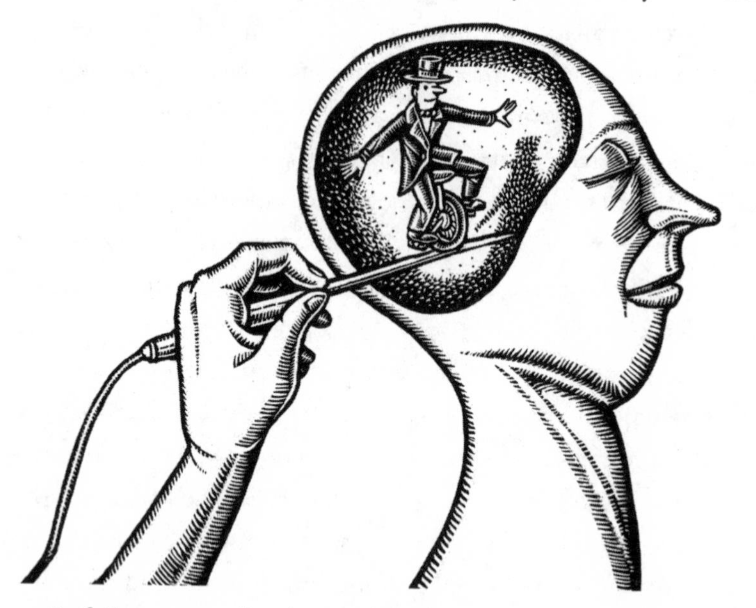

In follow-up studies, lasting benefits have been seen for as long as eight years after surgery, as long as progress has been tracked. The treatment incurs the same risk as all brain surgery: the small possibility of postoperative bleeding in the brain (see chapter 29). Although the benefits do decline with time, perhaps because Parkinson's disease continues to progress, patients almost always show long-term improvement. Plus, the new therapy allows patients to avoid the personality changes brought on by L-dopa treatment. The most common lasting side effect is a gain in weight averaging nine pounds, which is probably not a deterrent to those who need relief from their disease. By now, tens of thousands of patients have received stimulator implants. With these successes, it is no wonder that deep brain stimulation is a preferred treatment for advanced Parkinson's disease for anyone who can afford it or whose insurance covers it.

Still, despite the success of deep brain stimulation, we don't know exactly how it works. First, it's odd that stimulating a brain region would produce the same effect as a lesion. Stimulation probably doesn't kill brain tissue permanently, since the effects disappear when the treatment is stopped. One possible explanation is that stimulation has a jamming effect on whatever the subthalamic nucleus is trying to do. This could happen if stimulation interferes with impulses that would otherwise be generated in or pass through the subthalamic nucleus. Another possibility is that the high rate of stimulation reduces the amount of neurotransmitter available for release from subthalamic neurons, again reducing activity.

A second level of mystery is why blocking a signal from the subthalamic nucleus would help a parkinsonian brain initiate smooth movements at the right times. One guess is that the subthalamic nucleus's normal role is to oppose the substantia nigra's function. Removing its influence can then compensate for the loss of nigral function seen in Parkinson's patients. However stimulation of the brain works, the bottom line is that it allows high-level commands from the cortex to get through more clearly to the midbrain and spinal cord.

Deep brain stimulation treatment for Parkinson's disease has, in turn, led to other discoveries, often when doctors missed a surgical target, even by just a few millimeters. In one famous case, a woman was being treated for Parkinson's disease by deep brain stimulation. When her brain was tickled at a site just two millimeters (one-tenth of an inch) away from the place that relieved her motor symptoms, she became intensely depressed, weeping and saying things like "I'm disgusted with life . . . Everything is useless, always feeling worthless, I'm scared in this world." Fortunately, her symptoms disappeared about a minute after the stimulation ended. In other patients, stimulating another site, also just a few millimeters away, led to the opposite result: mania in the form of euphoria, nonstop talking, grandiose delusions, and increased sexual drive, all of which lasted for days. One of these patients asked repeatedly why he had not had the procedure done earlier. By the way, the answer to your question is no, you may not have this operation. Not yet, anyway.

An unavoidable impression from all the neurosurgical case studies reported to date is that we know very little about what many of these brain regions do. As we've pointed out before, structures such as the brainstem and midbrain are incredibly crowded, consisting of regions with very different functions piled next to one another cheek by jowl. Scientifically speaking, this can be considered a lucky accident, since surgeons' fortuitous discoveries in the middle of the brain would not be permitted as planned research.

In some cases, deep brain stimulation is starting to be applied in a rational fashion. For instance, surgery to treat obsessive-compulsives has focused on destroying a band of axons called the internal capsule, but a newer approach is to try deep brain stimulation at this location, a less damaging procedure. Another proposed therapy for depression is based on the observation that depressive episodes are associated with activity in a thin strip of cortical tissue called the subgenual cingulate, also known as area 25. Area 25 becomes less active in patients suffering from depression who respond to antidepressant drugs. In a small study, deep brain stimulation of the white matter underneath area 25 relieved symptoms in four of six patients with depression who could not be helped by medication, electroconvulsive therapy, or psychotherapy.

This approach to treating depression may eventually replace some extreme current treatments. The most effective therapy for major depression is electroconvulsive therapy, inducing

Did you know? Interfaces between brains and machines

In the classic novel *The Count of Monte Cristo*, Alexandre Dumas describes Monsieur Noirtier de Villefort, who after a stroke is alert and oriented to his surroundings, but is mute and paralyzed. He is able to communicate with others only by moving and blinking his eyes, and he conveys information using a list of letters. This disorder now has a name: locked-in syndrome. Locked-in people still have active brains but cannot translate their thoughts into actions. In addition to being caused by stroke, lock-in can result from neurological disorders, such as amyotrophic lateral sclerosis or ALS, which afflicts physicist Stephen Hawking. Spinal cord transection can also paralyze some or all of the limbs but spare speech, as happened to the late Christopher Reeve in a horseback-riding accident.

Researchers have been trying to design prosthetic mechanical limbs to help locked-in people gain some control over their surroundings. The idea is to monitor brain activity in the motor cortex to infer what movements patients are thinking of making. Such mind reading is possible, at least at a crude level, since even tetraplegics, who do not have control over any limb, show activity in the motor cortex when asked to think about movement. Arrays of electrodes can measure brain activity in a monkey as it moves its arm to play a video game, and researchers have used that activity to drive a mechanical arm. The resulting movements resemble those made by the monkey's own arm, albeit with a certain flailing quality and occasional moves in unexpected directions. Comparable progress has been made in an electrode array implanted into the brain of a human tetraplegic.

seizures throughout the entire brain, which can relieve symptoms for months (especially when paired with cognitive behavioral therapy). A therapy that is less extreme but less effective, and just about as mysterious, is stimulation of the vagus nerve, which helps one-third of persons suffering from depression who do not respond to antidepressant drugs. The vagus nerve conveys information to the brain about body systems, such as how fast the heart is beating, pain signals, and information from the gut and stomach (for instance, whether the stomach is full). One hypothesis is that this treatment works because feelings of well-being may depend on the interplay between body and brain signals. That is, vagus nerve stimulation may send happy-body signals to the brain.

Someday, deep brain stimulation may be designed rationally based on the known functions of the various parts of our brains. For the time being, though, we are limited by our basic knowledge about brain function. Scientists claim that deep brain stimulation is useful in treating problems like Tourette's syndrome and epilepsy, in addition to the disorders of movement and mood discussed above. It isn't clear whether deep brain stimulation can reliably help these patients, but that should become apparent if the treatment helps anyone as much as it helps parkinsonian patients. In the meantime, reports of weird effects of probing the brain's depths are a continuing source of evidence that when it comes to understanding how our brains work, we have a long way to go.

INDEX

Italic page numbers refer to illustrations.

A NOTE ON THE AUTHORS

Sandra Aamodt is the editor in chief of *Nature Neuroscience*, the leading scientific journal in the field of brain research. She received her undergraduate degree in biophysics from Johns Hopkins University and her doctorate in neuroscience from the University of Rochester. After four years of postdoctoral research at Yale University, she joined *Nature Neuroscience* at its founding in 1998 and became editor in chief in 2003. During her editorial career, she has read over three thousand neuroscience papers and written dozens of editorials on neuroscience and science policy for the journal. She has also given lectures at twenty universities, and attended forty-five scientific meetings in ten countries. She enjoys motorcycling and is preparing to spend a year sailing in the South Pacific. She lives with her husband, a professor of neuroscience, in California.

Sam Wang is an associate professor of neuroscience and molecular biology at Princeton University. He graduated with honor in physics from the California Institute of Technology at the age of nineteen and holds a doctorate in neuroscience from Stanford University School of Medicine. He has done research at Duke University Medical Center and at Bell Labs Lucent Technologies, and has worked on science and education policy for the United States Senate. He has published over forty articles on the brain in leading scientific journals, including *Nature*, *Nature Neuroscience*, *Proceedings of the National Academy of Sciences*, and *Neuron*. He is the recipient of a National Science Foundation Young Investigator Award and is an Alfred P. Sloan Fellow and a W. M. Keck Foundation Distinguished Young Scholar. He lives with his wife, a physician, and daughter in Princeton, New Jersey.